Design Guide for Concrete-Filled Double Skin Steel Tubular Structures

This is the first design guide on concrete-filled double skin steel tubular (CFDST) structures. It addresses in particular CFDST structures with plain concrete sandwiched between circular hollow sections and provides the relevant calculation methods and construction provisions for CFDST structures.

CFDST structures inherit the advantages of conventional concrete-filled steel tubular (CFST) structures, including high strength, good ductility and durability, high fire resistance and favorable constructability. Moreover, because of their unique sectional configuration, CFDST structures have been proved to possess lighter weight, higher bending stiffness and better cyclic performance than conventional CFST. Consequently, CFDST can offer reduced concrete consumption and construction costs.

This design guide is for engineers designing electrical grid infrastructures, wind power towers, bridge piers and other structures requiring light self-weight, high bending stiffness and high bearing capacity.

Design Guide for Concrete-Filled Double Skin Steel Tubular Structures

Lin-Hai Han
Dennis Lam
David A. Nethercot

CRC Press
Taylor & Francis Group
Boca Raton London New York

CRC Press is an imprint of the
Taylor & Francis Group, an **informa** business

CRC Press
Taylor & Francis Group
6000 Broken Sound Parkway NW, Suite 300
Boca Raton, FL 33487-2742

First issued in paperback 2021

ISBN-13: 978-1-138-34023-7 (hbk)
ISBN-13: 978-1-03-209449-6 (pbk)

Library of Congress Cataloging-in-Publication Data

Names: Han, Lin-Hai, author. | Lam, Dennis, author. | Nethercot, D. A., author.
Title: Design guide for concrete-filled double skin steel tubular structures / Lin-Hai Han, Dennis Lam and David Nethercot.
Description: First edition. | Boca Raton, FL : CRC Press/Taylor & Francis Group, [2019] | Includes bibliographical references and index.
Identifiers: LCCN 2018027397 (print) | LCCN 2018033839 (ebook) | ISBN 9780429802959 (Adobe PDF) | ISBN 9780429802942 (ePub) | ISBN 9780429802935 (Mobipocket) | ISBN 9781138340237 (hardback : acid-free paper) | ISBN 9780429440410 (ebook)
Subjects: LCSH: Tubular steel structures--Design and construction. | Concrete-filled tubes--Design and construction.
Classification: LCC TA684 (ebook) | LCC TA684 .H337 2019 (print) | DDC 624.1/821--dc23
LC record available at https://lccn.loc.gov/2018027397

Visit the Taylor & Francis website at
http://www.taylorandfrancis.com

and the CRC Press website at
http://www.crcpress.com

Contents

Preface

The concrete-filled double skin steel tubular (CFDST) structure is a new type of steel-concrete composite structure. It inherits the advantages of conventional concrete-filled steel tubular (CFST) structures, including high strength, good ductility and durability, high fire resistance and favorable constructability. Moreover, because of its unique sectional configuration, the CFDST structure has been proved to possess higher bending stiffness and better cyclic performance than conventional CFST structures. In addition, the use of CFDST structures could lead to lighter self-weight with less concrete consumption and a reduction of the foundation cost.

This design guide is based on the limit state design philosophy and is fully compliant with all the current Eurocodes. It covers the selection of materials, design of the CFDST structures, design of joints, fire protection requirements and practical construction issues. It will be a valuable guide for practicing engineers who are interested in the design of CFDST structures. Following the principle of sustainable development, the publication of this design guide is expected to exert positive effects on the application of CFDST structures.

Finally, the authors would like to thank and acknowledge the contributions from Mr. You-Xing Hua, Dr. Chao Hou, Dr. Wei Li, Dr. Kan Zhou and Dr. Therese Sheehan in preparation of this design guide.

Lin-Hai Han
Dennis Lam
David A. Nethercot

About the Authors

Lin-Hai Han: The research areas of Prof. Han are theoretical and experimental structural performance with particular expertise in steel-concrete composite and mixed structures. He teaches composite and mixed structures, structural fire safety design, disaster and disaster prevention to undergraduates and postgraduates. His research contributions have been recognized by a number of prestigious awards/prizes, including the Distinguished Young Scholar Award from the Natural Science Foundation of China. He served as Head of Department, Department of Civil Engineering, School of Civil Engineering, Tsinghua University from 2010 to 2016. He has been serving as the Associate Dean of the School since 2017. Professor Han was appointed as Professor of Chang Jiang Scholars Program of the Ministry of Education, P.R. China since 2014.

Dennis Lam: Chair of Structural Engineering and the Director of the Centre for Sustainable Environments at the University of Bradford, UK. He was Chief Structural Engineer for the City of Wakefield, UK, and is a fellow of the Institution of Structural Engineers. He is the President of Association of Steel-Concrete Composite Structures. He is also a member of the British Standard Institute and European Committee on Standardization (CEN), which is responsible for the Eurocode 4, and chairs the working group for the revision of the Eurocode 4 (EN1994-1-1).

David A. Nethercot: Former Head of the Department of Civil and Environmental Engineering at Imperial College, London. He was for more than 10 years chairman of the BSI Committee responsible for BS5950 and for UK input into EC3, is a past chairman of IABSE's technical committee, past Deputy Chairman of the Council of the Steel Construction Institute. He is Past President of the IStructE and a former Council Member of the Royal Academy of Engineering, 2008 recipient of the Charles Massonnet prize from the European Convention for Structural Steelwork and received a Gold Medal from the Institution of Structural Engineers in 2009.

CHAPTER 1

Introduction

1.1 CONCRETE-FILLED DOUBLE SKIN STEEL TUBES (CFDST)

Concrete-filled double skin steel tubular (CFDST) members are composite members which consist of inner skin, outer steel skin and concrete filled in the annulus. Figure 1.1 shows some typical CFDST cross sections. Either the inner skin or the outer skin can be circular, rectangular or square hollow sections. The outer steel skin can be either carbon steel or stainless steel. This type of composite structure derives from the conventional concrete-filled steel tube (CFST) structure and inherits the advantages of the CFST structure. This sandwiched cross section has been proved to have high stiffness with enhanced stability under external pressure. This section of the book introduces the applications of CFST structures briefly, and then presents an application of CFDST structures.

It is well known that the conventional CFST consists of an outer steel tube and a concrete core infilled. Various types of CFST cross sections are shown in Figure 1.2. Some background information is available in the publications by Zhao and Han (2006) and Han et al. (2014).

The advantages of CFST members, such as high strength and fire resistance, favorable ductility and high energy absorption

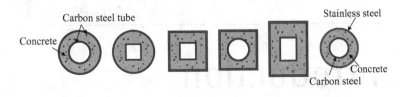

FIGURE 1.1 Typical CFDST cross sections.

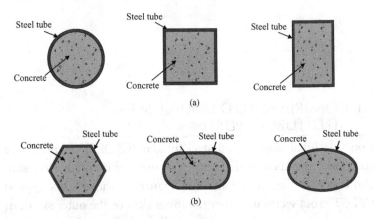

FIGURE 1.2 Typical CFST cross sections.

capacity, have been well recognized worldwide, (Goode and Lam, 2011; Lam and Williams, 2004; Qu et al., 2015; Yang et al., 2008) leading to the wide application of CFST members, especially in buildings, bridges and other large-scale structures. As shown in Figure 1.3, CFST members can be adapted to replace conventional

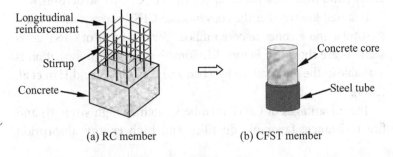

FIGURE 1.3 Typical RC and CFST members.

reinforced concrete (RC) members so that the structure can benefit from the advantages of CFST members. Moreover, the outer tube can be used as formwork in construction, significantly reducing the labor and material costs and increasing the efficiency of construction.

CFST structures have been used in China for over 50 years. They have been used as the load-bearing members in subway stations in Beijing since 1966 and in workshops and power plant buildings since the 1970s. Recent decades have seen the rapid development of the applications of CFST construction. CFST members are now being used as major load-bearing members under various loading conditions in buildings, bridges and other structures.

Figures 1.4 through 1.7 show the application of CFST members in an electrical transmission tower, a high-rise building and two bridges in China (Han, 2016; Han et al., 2014). The state-of-the-art research innovations for CFST structures and their development trends, especially in China, have been reviewed and discussed by Han et al. (2014).

The research and utilization of CFST structures in Europe are extensive as well. As shown in Figures 1.8 and 1.9, circular CFST

FIGURE 1.4   An electrical transmission tower in Zhoushan, China.

FIGURE 1.5   China Zun (528 m) under construction, Beijing, China.

columns are used in the Fleet Place House due to their good mechanical performance and appealing architectural appearance appealing architectural appearance (Hicks et al., 2002).

CFST columns are also gaining popularity in construction in Japan, where they are used as the critical columns in medium- and high-rise buildings, bridge piers, power transmission towers, and so on. The first edition of the AIJ standard for composite concrete and circular steel tubular structures was published in 1967, based on the research conducted in the early

FIGURE 1.6   Wushan Yangtze River Bridge, Chongqing, China.

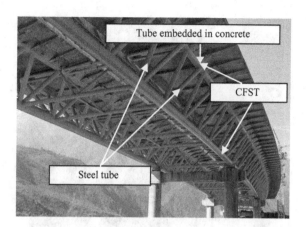

FIGURE 1.7 Ganhaizi Bridge, Yaan, Sichuan Province, China (1811 m in length with 36 spans and lattice piers up to 107 m in height).

1960s. Extensive work is also available on the investigation of CFST members and structures in the last two decades of the last century, including the "New Urban Housing Project" and the "US-Japan Cooperative Earthquake Research Program." In addition, individual universities and industrial institutions have also

FIGURE 1.8 Fleet Place House, London, UK.

FIGURE 1.9 Composite columns used in Fleet Place House, London, UK.

finished research work and presented the results at the annual meetings of the Architectural Institute of Japan (AIJ) (Morino and Tsuda, 2003).

The studies and application of CFST in the United States are relatively recent compared with those in Japan. The CFST members used in the United States differ significantly from those in Japan, which hindered the application of previous research in the United States. In Japan, the slenderness ratios of CFST members were typically not greater than 60, while in the United States, the slenderness ratios were greater than 100, with concrete used for maximizing the column stiffness (Roeder, 1998). When the CFST columns started to be used in United States, no relevant content was available in the design standards at that time. The application prompted experimental and theoretical research. For instance, Task Group 20 of the Structural Stability Research Council proposed a specification for the design of steel-concrete composite columns in 1979, which was subsequently adopted in the 1986 AISC-LRFD Code. Afterward, researchers studied the static and cyclic performance of CFST members and joints, and they proposed design methods to improve the practice of building CFST structures in the United States.

Most building structures with CFST columns used in the United States are braced frames, where axial forces are developed in these columns under lateral loading with a consequent demand for bond stress capacity. CFST columns are used in unbraced building structures as well, with the aim of forming primary lateral resistance systems. The building shown in Figure 1.10 is the Gateway Tower in Seattle, which employs a braced frame with CFST columns.

The application of the CFST structures is supported by several well-known national codes, such as the Japanese code AIJ (2001), American code AISC (American Institute of Steel Construction, 2010), British bridge code BS5400 (British Standards Institution, 2005), Australian bridge design standard AS5100 (Standards Australia, 2004), Chinese code DBJ/T13-51 (2010) and Eurocode 4 (2004).

The advantages of CFDST are similar to those of conventional CFST and include increased sectional stiffness; enhanced global stability; lighter weight; ability to use the space in the inner tube, if necessary, good damping characteristics and good cyclic performance. CFDST columns can exhibit enhanced fire resistance because the inner tubes are protected by the sandwiched concrete in the fire condition. Moreover, the steel tubes can act as formwork to the concrete during construction. The steel tubes can also bear a considerable amount of construction load prior to the pumping of the concrete mix, which results in construction efficiency.

FIGURE 1.10 Gateway Tower, Seattle, U.S.

CFDST has been used as high-rise bridge piers in Japan (Yagishita et al., 2000) to reduce the structural self-weight while maintaining a high energy-absorption capacity.

In recent years, many studies have been performed on CFDST members, and a state-of-the-art review has been conducted by Han et al. (2004) as well as Zhao and Han (2006). According to Han et al. (2004) and Tao et al. (2004), hollow ratio ψ is a critical parameter reflecting the proportion of the void in the cross section. This ratio is defined as $D_i/(D_o - 2t_o)$, where D_i and D_o are the diameters of the inner and outer tubes, respectively, and t_o is the

(a) lifting the outer tube

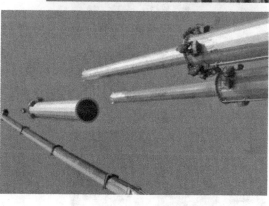

(b) placing the outer tube

FIGURE 1.11 A typical CFDST pole.

(c) pouring sandwiched concrete

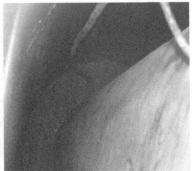

(d) fresh sandwiched concrete

(e) CFDST pole in service

FIGURE 1.11   (Continued)

thickness of the outer tube. If hollow ratio ψ of a CFDST column equals 0, this column is a conventional CFST column.

CFDST members have been used in electrical grid structures. Photos of a CFDST pole is shown in Figure 1.11. Figure 1.11a, b show both the outer and inner steel tubes of the CFDST during construction; Figure 1.11c, d show the process of concrete pouring between the gap of the outer and the inner steel tubes. Figure 1.11e shows the CFDST pole in service stage. The bearing capacity and stability of the pole is enhanced when compared with a conventional transmission tower, while the occupied area of the base is reduced.

1.2 ABOUT THIS GUIDE

The design guide is intended to be used by designers and engineers for the design, construction and inspection of this form of structure and is most suitable for those CFDST structures with plain concrete sandwiched between circular hollow sections (CHS). It discusses the selection of materials, design of the CFDST structures, design of joints, fire protection requirements and practical construction issues for CFDST structures.

1.3 DEFINITIONS

1.3.1 Concrete-Filled Double Skin Steel Tubular Structure

Structure consisting of inner and outer steel skins with the annulus between the skins filled with concrete.

1.3.2 Hollow Ratio

Given by $D_i/(D_o - 2t_o)$, where t_o is the wall thickness of the outer steel tube, D_i and D_o are the outer diameter of the inner tube and the outer width or diameter of the outer tube, respectively.

1.3.3 Steel Ratio

Given by A_{so}/A_c, where A_{so} and A_c are the cross-sectional area of the outer steel tube and the cross-sectional area of the concrete, respectively.

1.3.4 Nominal Steel Ratio

Given by A_{so}/A_{ce}, where A_{so} and A_{ce} are the cross-sectional area of the outer steel tube and the nominal cross-sectional area of the concrete (given by $A_{ce} = (\pi/4)(D_o - 2t_o)^2$), respectively.

1.3.5 Nominal Confinement Factor

Given by $\alpha_n f_{y,outer}/f_{ck}$, where α_n is the nominal steel ratio of the CFDST columns, $f_{y,outer}$ is the yield stress of the outer steel tube and f_{ck} is the characteristic compressive strength of the concrete.

1.4 SYMBOLS

1.4.1 Design Action and Action Effect

M	Bending moment
M_c	Design moment at the beam end
N	Axial force
N_{LF}	Design value of bearing capacity for CFDST members to resist lateral local compression
V	Shear force

1.4.2 Computational Indices

E_c	Modulus of elasticity of concrete
E_s	Modulus of elasticity of steel
$(EA)_{sc}$	Compound elastic compression stiffness of concrete-filled double skin steel tube
$(EI)_{sc}$	Compound elastic flexural stiffness of concrete-filled double skin steel tube
G_c	Shear modulus of elasticity of concrete
G_s	Shear modulus of elasticity of steel
$(GA)_{sc}$	Compound elastic shear stiffness of concrete-filled double skin steel tube
N_b	Tension of the ring plate caused by axial force of the beam
$N_{osc,u}$	Sectional capacity of the outer tube and the sandwiched concrete
$N_{i,u}$	Sectional capacity of the inner tube

N_E Elastic bucking load

N_{LF} The design value of the lateral local compression in the joint area of CFDST members

N_u Sectional capacity of the CFDST members under axial compression

N_{us} The bearing capacity of the member under short-term load

N_{uLF} Design value of bearing capacity for CFDST members to resist lateral local compression

N_{ut} Sectional capacity of the CFDST tensile member

ΔN_{ic} Axial load sustained by sandwich concrete transmitted from the beam of floor i which is connected to the column

M_u Sectional flexural bearing capacity

$M_{osc,u}$ Compound flexural bearing capacity of the outer tube and sandwiched concrete

$M_{i,u}$ Flexural bearing capacity of the inner tube

$N_{x,max}$ The tension caused by the combination of the most adverse effects in X direction

N_y The tension in Y direction (working with $N_{x,max}$ at the same time)

W_{scm} Compound flexural modulus of the outer tube and sandwiched concrete

W_{si} The flexural modulus of the inner tube

f_1 Design value of tensile strength of the steel of the stiffening ring plate

f_b Design value of the bond strength between steel tube and concrete, and for the bonding between circular steel tube and concrete

f_{ck}, f_c Characteristic value and design value of compressive strength of concrete

f_{tk}, f_t Characteristic value and design value of tensile strength of concrete

f_y Yield strength of steel

$f_{y,inner}$ Yield strength of steel for inner tube

$f_{y,outer}$ Yield strength of steel for outer tube

f Design value of tensile, compressive and flexural strength for steel, and design value of tensile strength of the steel bar in the section with negative bending moment at the end of the beam

f_{inner} Design value of tensile, compressive and flexural strength for the steel of the inner tube

f_{outer} Design value of tensile, compressive and flexural strength for the steel of the outer tube

f_{osc} Design value of compound axial compressive strength for the outer tube and the sandwiched concrete

f_{jv} Design value of shear strength of the weld

1.4.3 Geometric Parameter

A_c Cross-sectional area of sandwiched concrete

A_{ce} Nominal cross-sectional area of concrete in CFDST

A_s Area of the steel bar in the section with negative bending moment at the end of the beam

A_{sc} Gross cross-sectional area of the composite member

A_{si} Area of inner steel hollow section in CFDST

A_{so} Area of outer steel hollow section in CFDST

A_{soe} Equivalent sectional area of outer tube

A_1 Area of local compression for CFDST (For hollow branch pipe, the area is the same value with solid branch pipe which has identical diameter), $A_1 = \dfrac{\pi d_w^2}{4}$, d_w is diameter of the brace which transfers lateral forces

A_b Maximum design distribution area for lateral partially compression

A_d Sectional area of the oblique web tube in lattice CFDST member

C Section perimeter of the outer tube in CFDST

D_o Outer diameter of the outer tube in CFDST

D_i Outer diameter of the inner tube in CFDST

D_t Diameter of outer circular steel tube for the top cross section of concrete-filled double skin steel tubular tapered column

D_b Diameter of outer circular steel tube for the bottom cross section of concrete-filled double skin steel tubular tapered column

D_e Equivalent external diameter of outer tube

I_c Second moment of area of concrete

I_{si} Second moment of area of inner tube

I_{so} Second moment of area of outer tube

l_0 Effective length of a member

l_{t0} Effective length of a tapered member

l Length between the middle points of the top and bottom floor column

l_1 Length of chord in an interval

l_c Perimeter of inner surface section of outer tube or outer surface section of inner tube

l_j Length of weld

b Halfwidth of the quadruple and double component column

b_1, b_2 Distance between the centroidal axis of the cross section and the centroidal axis of each column component in the wide direction

b_e Effective width of the outer tube wall of the column participating in the work of the stiffening ring

b_j Width filled by fillet weld

b_s Width of the stiffening ring plate

d_c Distance between concrete and tube along radial direction in circumferential gap

d_s Distance between concrete and tube along radial direction in circular-segment gap

e Load eccentricity

h Height of the section of the end of the beam; half height of the quadruple and triple component column; the depth which the column plug into the base cup

h_f Weld height

h_j Height of the web or the plate

r_c Size of the nominal core concrete section $r_c = D_o/2 - t_o$

t Fire resistance, in h

t_1	Thickness of the stiffening ring plate
t_i	Wall thickness of the inner steel tube in CFDST
t_o	Wall thickness of the outer steel tube in CFDST
Δt_o	Thickness loss of outer tube under corrosion
t_{oe}	Equivalent wall thickness of outer tube
t_w	Thickness of the weld (plate)
ψ_c	Ratio of circumferential gap
ψ_s	Ratio of circular-segment gap
γ	Taper ratio
θ	Tapered angle

1.4.4 Computational Coefficients

n	Axial compression ratio, long-term load ratio
k_{cr}	Long-term load influence coefficient, referred to Table 3.2
k_t	Strength index for CFDST columns
α	Steel ratio
α_1	Sectional area ratio of the outer tube of the single chord to the web tube
α_b	Sectional area ratio of the outer tube of the single chord to the horizontal web tube
α_d	Sectional area ratio of the outer tube of the single chord to the oblique web tube
α_n	Nominal steel ratio; nominal steel ratio of a single chord
α_{ne}	Equivalent nominal steel ratio
α_{io}	Sectional area ratio of the inner tube to the outer tube of a single chord
α_E	Modular ratio of steel to concrete $\alpha_E = E_s / E_c$
β_l	Strengthen coefficient for the concrete supposed to lateral local compression
β_c	Influence coefficient for concrete strength subjected to lateral local compression
β_h	Ratio of the vertical biaxial tension of the stiffening ring plate
β_m	Equivalent moment factor
ψ	Hollow section ratio

ψ_e Equivalent hollow section ratio

γ_{m1}, γ_{m2} Coefficient of cross-sectional moment capacity

χ Stability factor of axial compression member

λ Slenderness ratio

λ_1 Slenderness ratio of a single chord in an interval

λ_o Slenderness ratio of the horizontal web members

λ_{max} The larger of the equivalent slenderness ratio in the X-X and Y-Y direction

τ_s Bond stress between steel tube and sandwich concrete

ξ Nominal confinement factor

ξ_e Equivalent standard value of nominal confinement factor

ξ_o Design confinement factor

CHAPTER 2

Materials

2.1 STEEL

2.1.1 The properties should be obtained in accordance with EN 1993-1-1 (Sections 3.1 and 3.2). The quality requirements should comply with the corresponding product standards. Other types of steel should also comply with relevant product standards if used.

The strength classes in this design guide are based on the yield strength f_y with a maximum value of S_{max}.

Note: The value of S_{max} for application within this design guide is 460 N/mm². Steel of grades S235, S275 and S355 are suggested. Use of high strength steel beyond the current limit required further consideration.

2.1.2 For the circular CFDST tube, straight seam welded tube and spiral welded tube will be used.

2.2 CONCRETE

2.2.1 The sandwiched concrete can be normal concrete or high-strength concrete; the water-cement ratio should not be greater than 0.45. The characteristic (5%) cylinder strength f_{ck} of the sandwiched concrete should be within the range from 25 N/mm² to 56 N/mm²; the properties and quality requirement should comply with EN 1992-1-1 Section 3.1 and EN 1990 Section 2. Use of high strength concrete beyond the current limit required further consideration.

2.2.2 Normally the strength of the sandwiched concrete should be compatible with the strength of the steel tube, that is, concrete of higher strength class should be used with steel tube of higher strength grade.

2.2.3 The shrinkage strain value of sandwiched concrete can be calculated using Appendix 1.

2.3 CONNECTION DEVICES

2.3.1 Fasteners and welding consumables

Reference should be made to EN 1993-1-8, Chapters 3 and 4 for requirements for fasteners and welding consumables.

2.3.2 Headed stud shear connectors

Reference should be made to EN 13918.

CHAPTER 3

Design of CFDST Columns

3.1 DESIGN PHILOSOPHY

3.1.1 General Rules

General rules for the design of CFDST structures are presented herein, including rules for the limitation of parameters in current equations.

1. The structural design of CFDST must be on the basis of national construction principles and technical economy policy, the engineering condition, material supply, transport, setup and construction of members should be fully considered. The structural plan needs to be designed reasonably to confirm that it is safe, applicable, economical, resource conservative and environmentally friendly.

2. Ultimate strength design methods based on probability theory are adopted, the design resistance is calculated using the equations with partial factors.

3. The CFDST structures need to meet the ultimate limit state and the serviceability limit state requirements.

a. The strength limit state should consider the basic combination of load effect while the accidental combination should also be considered if necessary. The bearing capability of members should be calculated with the design value of load. For structures with requirements of seismic resistance, seismic checking based on EN 1998-1: 2004 and other relevant standards are needed as well.

b. The design for serviceability limit state should consider the standard combination of load effects.

4. The seismic design of CFDST structures should meet the provisions of EN 1998-1: 2004.

5. The diameter of the outer tube should not be less than 200 mm, the thickness of the tube should not be less than 4 mm and the maximum aggregate size of the concrete should not exceed 1/3 of the clear distance between the inner and outer tubes. The ratio of the diameter of the outer tube to the wall thickness should not exceed 1.5 times the limit value of the hollow tube without concrete infill. This limit value without concrete infill is obtained by:

$$\frac{D_o}{t_o} \leq 135\varepsilon^2 \tag{3.1}$$

where $\varepsilon = \sqrt{235/f_y}$

The ratio of diameter to thickness of inner CHS should meet the provisions for hollow tubes under axial compression in EN 1993-1-1 (Table 5.2).

6. The hollow section ratio of CFDST should be kept between 0 and 0.75. The hollow section ratio should be calculated as follows:

$$\psi = \frac{D_i}{D_o - 2t_o} \tag{3.2}$$

where:

ψ is the hollow section ratio of CFDST

D_i is the outside diameter of inner CHS in CFDST

D_o is the outside diameter of outer CHS in CFDST

t_o is the wall thickness of the outer steel tube in CFDST

7. When CFDST is used in seismically active areas, the nominal confinement factor of circular CFDST members should not be less than 0.6 or greater than 4.0. The nominal confinement factor should be calculated as follows:

$$\xi = \frac{A_{so} \cdot f_{y,\text{outer}}}{A_{ce} \cdot f_{ck}} = \alpha_n \cdot \frac{f_{y,\text{outer}}}{f_{ck}} \tag{3.3}$$

where:

α_n is the nominal steel ratio

A_{so} is the cross-sectional area of outer steel hollow section in CFDST

A_{ce} is the nominal cross-sectional area of concrete in CFDST, $A_{ce} = \pi(D_o - 2t_o)^2 / 4$

$f_{y,\text{outer}}$ is the yield strength of steel of the outer tube

f_{ck} is the characteristic compressive strength of concrete

8. Prefabricated parts should be checked for construction safety. When checking the lifting for prefabricated parts, the dead load should be multiplied by an amplification factor of 1.5.

9. The separation between the concrete and steel tube should be measured. The provisions in Sections 3.9.3 and 3.9.4 should be used to confirm if the strength reduction caused by the circumferential gap or the circular-segment gap needs to be considered in design. If necessary, the calculation of the

bearing capacity for CFDST members should take into consideration the effect of imperfection according to Section 3.9.

3.1.2 Design Indices

1. The nominal value of the strength for hot rolled steel tubes or the steel tubes made of hot rolled steel plate can be obtained from EN 1993-1-1 (Table 3.1) or relevant product standards (EN 10025-2).

2. The nominal value of the strength for thin-walled cold-formed steel tubes can be obtained from EN 1993-1-3 (Table 3.1a and 3.1b).

3. The material properties of steel can be obtained from EN 1993-1-1 (Section 3.2) or product standards (EN 10025-2).

4. The design value, characteristic value and modulus of elasticity of concrete can be obtained from EN 1992-1-1 (Section 3.1).

5. The compound elastic compression stiffness of CFDST should be calculated as follows:

$$(EA)_{sc} = E_s \cdot (A_{si} + A_{so}) + E_c \cdot A_c \qquad (3.4)$$

where:

E_s, E_c are the modulus of elasticity of steel and concrete

A_{si}, A_{so} are the cross-sectional area of inner and outer tube

A_c is the cross-sectional area of sandwiched concrete

6. Compound elastic flexural stiffness of CFDST should be calculated as follows:

$$(EI)_{sc} = E_s \cdot (I_{si} + I_{so}) + a \cdot E_c \cdot I_c \qquad (3.5)$$

where:

I_{si}, I_{sc}, I_c are the moment of inertia of the cross sections of inner tube, outer tube and sandwiched concrete, respectively

a is the stiffness reduction coefficient of sandwiched concrete, $a = 0.6$

7. Compound elastic shear stiffness of CFDST should be calculated as follows:

$$(GA)_{sc} = G_s \cdot (A_{si} + A_{so}) + G_c \cdot A_c \qquad (3.6)$$

where:

G_s is the shear modulus of elasticity of steel

G_c is the shear modulus of elasticity of concrete, 0.4 times the value of modulus of elasticity of concrete.

3.1.3 Deformation of Structures and Members

1. When CFDST is used in power transmission tower structures, the limitation of its deformation should satisfy the stipulation of the electricity standard EN 1998-6 (Section 5.5).

2. When conducting static or dynamic analysis, the structure should be considered as a whole part with a proper calculation model adopted.

3.1.4 Anti-Corrosion

1. When the structure is exposed to the environments that have special requirements for corrosion prevention or is subjected to corrosive gaseous and solid mediums, weathering steel should be used.

2. Anti-corrosion measures should be taken in order to protect the outer tube for CFDST, such as paint or the use of metal galvanizing after applying anti-rust protection. The surface protection should be obtained in accordance with EN ISO 12944.

3.2 CROSS-SECTIONAL RESISTANCE

3.2.1 The axial resistance of a cross section for CFDST members should meet the following requirement:

a. Axial compression

The cross-sectional resistance of axial compressive CFDST members should be determined as follows:

$$N_u = N_{osc,u} + N_{i,u} \tag{3.7}$$

$$N_{osc,u} = f_{osc} \cdot (A_{so} + A_c) \tag{3.8}$$

$$N_{i,u} = f_{inner} \cdot A_{si} \tag{3.9}$$

where:

N_u is the sectional capacity of the CFDST members under axial compression

$N_{osc,u}$ is the sectional capacity of the outer tube and the sandwiched concrete

$N_{i,u}$ is the sectional capacity of the inner tube

f_{osc} is the design value of compound axial compressive strength for the outer tube and the sandwiched concrete

f_{inner} is the design value of tensile, compressive and flexural strength for the steel of the inner tube

A_{si} is the sectional area of the inner tube

The composite strength of the outer tube and the sandwiched concrete f_{osc} should be obtained from Equation (3.10).

$$f_{osc} = C_1 \cdot \psi^2 \cdot f_{outer} + C_2 \cdot (1.14 + 1.02\xi_o) \cdot f_c \qquad (3.10)$$

where:

C_1 is a calculation coefficient, $C_1 = \alpha/(1+\alpha)$

C_2 is a calculation coefficient, $C_2 = (1+\alpha_n)/(1+\alpha)$

ξ_o is the design confinement factor, $\xi_o = \dfrac{A_{so} \cdot f_{outer}}{A_{ce} \cdot f_c} = \alpha_n \cdot \dfrac{f_{outer}}{f_c}$

α is the steel ratio of the member, $\alpha = A_{so}/A_c$

α_n is the nominal steel ratio of the member, $\alpha_n = A_{so}/A_{ce}$

ψ is the hollow section ratio of CFDST

f_{outer} is the design value of tensile, compressive and flexural strength for the steel of the outer tube

f_c is the design value of concrete strength

b. Axial tension

The cross-sectional resistance of axial tensile CFDST members should be calculated as follows:

$$N_{ut} = (1.1 - 0.05\xi_o) \cdot (1.09 - 1.06\psi) \cdot f_{outer} \cdot A_{so} + f_{inner} \cdot A_{si} \qquad (3.11)$$

3.2.2 The cross-sectional flexural resistance of CFDST members should meet the following requirement:

$$M_u = M_{osc,u} + M_{i,u} \qquad (3.12)$$

where:

M_u is the flexural bearing capacity

$M_{osc,u}$ is the compound flexural bearing capacity of the outer tube and sandwiched concrete

$M_{i,u}$ is the flexural bearing capacity of the inner tube

$$M_{osc,u} = \gamma_{m1} \cdot W_{scm} \cdot f_{osc} \qquad (3.13)$$

$$M_{i,u} = \gamma_{m2} \cdot W_{si} \cdot f_{inner} \qquad (3.14)$$

where:

W_{scm} is the compound flexural modulus of the outer tube and sandwiched concrete, $W_{scm} = \dfrac{\pi(D_o^4 - D_i^4)}{32D_o}$

W_{si} is the flexural modulus of the inner tube, $W_{si} = \dfrac{\pi\left[D_i^4 - (D_i - 2t_i)^4 \right]}{32D_i}$

γ_{m1}, γ_{m2} are the flexural bearing capacity factor

$$\gamma_{m1} = 0.48\ln(\xi + 0.1) \cdot (1 + 0.06\psi - 0.85\psi^2) + 1.1 \qquad (3.15)$$

$$\gamma_{m2} = -0.02\psi^{-2.76}\ln\xi + 1.04\psi^{-0.67} \qquad (3.16)$$

The selected cross section for the calculation of flexural bearing capacity for the tapered CFDST member should be the minimum cross section.

3.3 RESISTANCE OF MEMBERS UNDER AXIAL LOADING

3.3.1 The effective length of the CFDST column should be determined according to the regulations in EN 1993-1-1 (Annex B.1).

3.3.2 The slenderness limit of the CFDST column should be determined according to the regulations in EN 1993-1-1 (Chapter 6).

3.3.3 The bearing capacity of the CFDST members under axial loading should meet the following requirements:

a. Axial compression

$$N \leq \chi \cdot N_u \qquad (3.17)$$

where:

N is the design value of the axial compressive load applied on the CFDST member

χ is the stability reduction factor of the axial compression member (Appendix 2, where λ is the slenderness ratio of the member)

N_u is the sectional capacity of the CFDST members under axial compression

The slenderness ratio of the member should be calculated as follows:

$$\lambda = \frac{4l_0}{\sqrt{D_o^2 + (D_i - 2t_i)^2}} \qquad (3.18)$$

For the tapered CFDST members with hinged ends, the minimum cross section should be considered for the sectional capacity of the member and the effective length should be calculated as follows:

$$l_{t0} = \frac{l_0}{\sqrt{2\gamma + 1}} \qquad (3.19)$$

where:

l_0 is the effective length of the corresponding constant cross section column which has the same geometric length and boundary condition

γ is the taper ratio

$$\gamma = \frac{D_b - D_t}{D_t} \qquad (3.20)$$

where:

D_t is the diameter of the outer circular steel tube for the top cross section of the CFDST tapered column

D_b is the diameter of the outer circular steel tube for the bottom cross section of the CFDST tapered column

1. Axial tension:

$$N < N_{ut} \tag{3.21}$$

where:

N_{ut} is the sectional capacity of the CFDST tensile member

N_{ut} should be obtained according to provision 3.2.1.

3.3.4 The buckling resistance of the whole member and each individual component of the lattice compressive CFDST members should be checked.

3.3.5 The buckling resistance of the whole lattice compressive CFDST member should be calculated using Equation (3.17), the stability factor can be found in Appendix 2 based on the equivalent slenderness ratio, which is given by Table 3.1.

In Table 3.1:

l_1 is length of chord in an interval

λ_1 is the slenderness ratio of a single chord in an interval

λ_o is the slenderness ratio of the horizontal web members

λ_x, λ_y are the slenderness ratios of the whole member about axes X-X and Y-Y, respectively

α_E is the modular ratio of steel to concrete, $\alpha_E = E_s/E_c$

α_n is the nominal steel ratio of a single chord, $\alpha_n = A_{so}/A_{ce}$

α_1 is the sectional area ratio of the outer tube of the single chord to the web tube, $\alpha_1 = A_{so}/A_1$

α_d is the sectional area ratio of the outer tube of the single chord to the oblique web tube, $\alpha_d = A_{so}/A_d$

α_b is the sectional area ratio of the outer tube of the single chord to the horizontal web tube, $\alpha_b = A_{so}/A_b$

α_{io} is the sectional area ratio of the inner tube to the outer tube of a single chord, $\alpha_{io} = A_{si}/A_{so}$

b is the half width of the quadruple and double component column

TABLE 3.1 The Equivalent Slenderness Ratio of Lattice Members

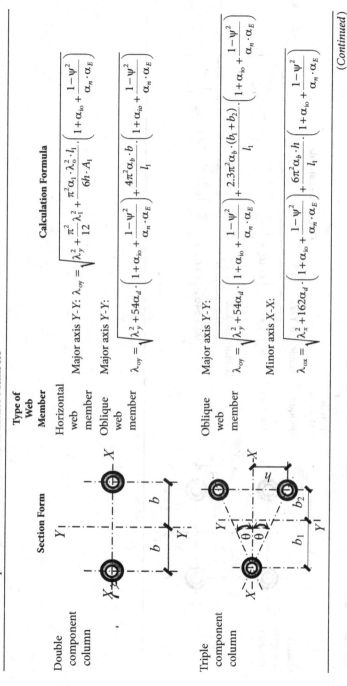

Section Form	Type of Web Member	Calculation Formula
Double component column	Horizontal web member	Major axis Y-Y: $\lambda_{oy} = \sqrt{\lambda_y^2 + \dfrac{\pi^2}{12}\lambda_1^2 + \dfrac{\pi^2\alpha_1\cdot\lambda_o^2\cdot l_1}{6h\cdot A_1}\cdot\left(1+\alpha_{io}+\dfrac{1-\psi^2}{\alpha_n\cdot\alpha_E}\right)}$
	Oblique web member	Major axis Y-Y: $\lambda_{oy} = \sqrt{\lambda_y^2 + 54\alpha_d\cdot\left(1+\alpha_{io}+\dfrac{1-\psi^2}{\alpha_n\cdot\alpha_E}\right)+\dfrac{4\pi^2\alpha_b\cdot b}{l_1}\cdot\left(1+\alpha_{io}+\dfrac{1-\psi^2}{\alpha_n\cdot\alpha_E}\right)}$
Triple component column	Oblique web member	Major axis Y-Y: $\lambda_{oy} = \sqrt{\lambda_y^2 + 54\alpha_d\cdot\left(1+\alpha_{io}+\dfrac{1-\psi^2}{\alpha_n\cdot\alpha_E}\right)+\dfrac{2.3\pi^2\alpha_b\cdot(b_1+b_2)}{l_1}\cdot\left(1+\alpha_{io}+\dfrac{1-\psi^2}{\alpha_n\cdot\alpha_E}\right)}$
		Minor axis X-X: $\lambda_{ox} = \sqrt{\lambda_x^2 + 162\alpha_d\cdot\left(1+\alpha_{io}+\dfrac{1-\psi^2}{\alpha_n\cdot\alpha_E}\right)+\dfrac{6\pi^2\alpha_b\cdot h}{l_1}\cdot\left(1+\alpha_{io}+\dfrac{1-\psi^2}{\alpha_n\cdot\alpha_E}\right)}$

(Continued)

TABLE 3.1 (CONTINUED) The Equivalent Slenderness Ratio of Lattice Members

Section Form	Type of Web Member	Calculation Formula
Quadruple component column 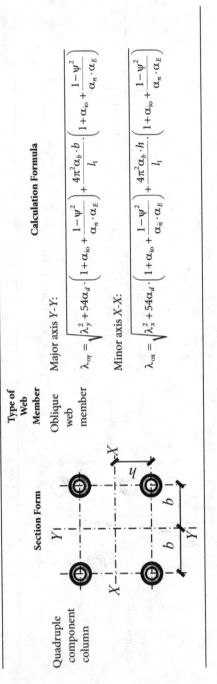	Oblique web member	Major axis Y-Y: $\lambda_{oy} = \sqrt{\lambda_y^2 + 54\alpha_d \cdot \left[1 + \alpha_{io} + \dfrac{1-\psi^2}{\alpha_n \cdot \alpha_E}\right] + \dfrac{4\pi^2 \alpha_b \cdot b}{l_1} \cdot \left(1 + \alpha_{io} + \dfrac{1-\psi^2}{\alpha_n \cdot \alpha_E}\right)}$ Minor axis X-X: $\lambda_{ox} = \sqrt{\lambda_x^2 + 54\alpha_d \cdot \left[1 + \alpha_{io} + \dfrac{1-\psi^2}{\alpha_{\bar{n}} \cdot \alpha_E}\right] + \dfrac{4\pi^2 \alpha_b \cdot h}{l_1} \cdot \left(1 + \alpha_{io} + \dfrac{1-\psi^2}{\alpha_n \cdot \alpha_E}\right)}$

h is the half height of the quadruple and triple component column

b_1, b_2 are the distance between the centroidal axis of the cross section and the centroidal axis of each column component in the wide direction, respectively

3.3.6 The buckling resistance of a circular lattice compressive CFDST member should be checked using Equation (3.17), and the buckling resistance of a single component column should also be checked. When the slenderness ratio λ_1 of an individual component column satisfies the following conditions, its buckling resistance does not need to be checked:

For the lattice member with horizontal web members: $\lambda_1 \leq 40$ and $\lambda_1 \leq 0.5\lambda_{max}$; for the lattice member with oblique web members: $\lambda_1 \leq 0.7\lambda_{max}$;

where:

λ_{max} is the larger of the equivalent slenderness ratios about axes X-X and Y-Y

3.4 RESISTANCE OF MEMBERS IN COMBINED AXIAL LOADING AND BENDING

3.4.1 The bearing capacity of CFDST members subjected to compression and bending within a plane should meet the following requirements:

a. when $N/N_u \geq 2\chi^3\eta_o$:

$$\frac{N}{\chi \cdot N_u} + \frac{a}{d} \cdot \left(\frac{\beta_m \cdot M}{M_u} \right) \leq 1 \qquad (3.22)$$

b. when $N/N_u < 2\chi^3\eta_o$:

$$-b\left(\frac{N}{N_u} \right)^2 - c\left(\frac{N}{N_u} \right) + \frac{1}{d} \cdot \left(\frac{\beta_m \cdot M}{M_u} \right) \leq 1 \qquad (3.23)$$

where:

$$a = 1 - 2\chi^2 \cdot \eta_o; \quad b = \frac{1-\zeta_o}{\chi^3 \cdot \eta_o^2}; \quad c = \frac{2(\zeta_o - 1)}{\eta_o}; \quad d = 1 - 0.4\left(\frac{N}{N_E}\right);$$

$$N_E = \frac{\pi^2 (E_s I_{so} + E_s I_{si} + E_c I_c)}{l_0^2};$$

For circular CFDST members:

$$\zeta_o = 1 + (0.18 - 0.2\psi^2)\xi^{-1.15}$$

$$\eta_o = \begin{cases} (0.5 - 0.245\xi) \cdot (1 + 0.7\psi - 1.8\psi^2) & \xi \leq 0.4 \\ (0.1 + 0.14\xi^{-0.84}) \cdot (1 + 0.7\psi - 1.8\psi^2) & \xi > 0.4 \end{cases}$$

M is the maximum design value of the bending moment in the calculated section of the members

β_m is the equivalent moment factor, which can be obtained in EN 1994-1-1 (Table 6.4)

L_e is the effective length of the members

N_u is the sectional strength of the members under axial compression, which should be calculated using Equation (3.7)

M_u is the sectional flexural bearing capacity, which should be calculated using Equation (3.12)

The effective length of the tapered CFDST beam-column members should be calculated using Equation (3.19), and the calculated cross section for compression and flexural bearing capacity of the tapered CFDST member should be the minimum cross section.

3.4.2 The bearing capacity of the CFDST members under bending and tension should meet the following requirement:

$$\frac{N}{N_{ut}} + \frac{M}{M_u} \leq 1 \tag{3.24}$$

where:

N is the design value of the tensile load

N_{ut} is the sectional strength of the member under axial tension, which should be calculated as Equation (3.11)

M is the maximum design value of the bending moment in the calculated section of the member

M_u is the sectional flexural bearing capacity, which should be calculated as Equation (3.12)

3.4.3 When lattice CFDST members are subjected to compression and bending, the in-plane overall buckling resistance should be checked as follows:

$$\frac{N}{\chi \cdot N_u} + \frac{\beta_m \cdot M}{W_{sc} \cdot (1 - \chi \cdot N/N_E) \cdot f_{osc}} \leq 1 \qquad (3.25)$$

$$N_E = \frac{\pi^2 (EA)_{sc}}{\lambda^2} \qquad (3.26)$$

where:

χ is the stability factor for the check of in-plane stability of the axial compression member (refer to Appendix 2)

A_{sc}, W_{sc} are the whole area and section modulus of lattice members

λ is the equivalent slenderness ratio

β_m is the equivalent moment factor, which can be obtained in EN 1994-1-1 (Table 6.4)

It is not necessary to check the out-of-plane overall stable bearing capacity, but the stability of the component column should be checked. For the single component column of the lattice column with oblique web members, it can be treated as a chord member in the truss. For the single component of the lattice column with horizontal web members, it should be treated as an eccentrically compressed member, subjected to local moment caused by shear.

3.5 LONG-TERM EFFECTS

3.5.1 The bearing capacity of CFDST members under long-term load should meet the following requirement:

$$N \le k_{cr} \cdot N_{us} \tag{3.27}$$

where:

k_{cr} is the long-term load influence coefficient, refer to Table 3.2

N_{us} is the bearing capacity of the member under short-term load, which can be calculated by Equations (3.7), (3.22), (3.23), (3.24) and (3.25)

In Table 3.2,

n is the long-term load ratio, $n = N/N_{us}$

N is the long-term load

e is the load eccentricity

r is the outside radius of the outer steel tube in CFDST, $r = D_o / 2$

3.6 LOCAL BEARING IN THE CONNECTION AREA

3.6.1 For CFDST members under lateral local bearing in the connection areas, as shown in Figure 3.1, to avoid possible local bearing failure, the following requirements should be satisfied:

a. The dimension of the brace should not be greater than that of the chord;

b. The wall thickness of the brace should not be greater than that of the chord;

c. Under loading, the brace should not penetrate the chord.

TABLE 3.2 The Long-Term Load Influence Coefficient k_{cr} of Circular CFDST Member

Hollow Section Ratio ψ	Nominal Confinement Factor ξ	Eccentricity Ratio e/r	Long-Term Load Ratio n	Slenderness Ratio λ					
				20	40	80	100	120	160
0	1	0	0.2	0.915	0.860	0.813	0.813	0.829	0.845
			0.5	0.896	0.841	0.796	0.796	0.812	0.828
			0.8	0.876	0.823	0.779	0.779	0.795	0.811
		0.5	0.2	0.946	0.917	0.854	0.848	0.859	0.867
			0.5	0.925	0.898	0.836	0.830	0.841	0.850
			0.8	0.905	0.878	0.818	0.812	0.824	0.832
		1	0.2	0.956	0.937	0.868	0.860	0.869	0.875
			0.5	0.936	0.917	0.850	0.842	0.852	0.857
			0.8	0.915	0.897	0.832	0.824	0.834	0.839
	2	0	0.2	0.931	0.890	0.842	0.841	0.858	0.875
			0.5	0.911	0.871	0.824	0.824	0.841	0.857
			0.8	0.891	0.852	0.806	0.807	0.823	0.840
		0.5	0.2	0.962	0.949	0.884	0.877	0.889	0.897
			0.5	0.942	0.929	0.865	0.859	0.871	0.879
			0.8	0.921	0.909	0.847	0.841	0.853	0.861
		1	0.2	0.973	0.970	0.899	0.890	0.900	0.905
			0.5	0.952	0.950	0.880	0.872	0.882	0.887
			0.8	0.931	0.929	0.861	0.853	0.863	0.869

(Continued)

TABLE 3.2 (CONTINUED) The Long-Term Load Influence Coefficient kcr of Circular CFDST Member

Hollow Section Ratio ψ	Nominal Confinement Factor ξ	Eccentricity Ratio e/r	Long-Term Load Ratio n	Slenderness Ratio λ					
				20	40	80	100	120	160
	3	0	0.2	0.940	0.908	0.859	0.859	0.876	0.893
			0.5	0.920	0.889	0.841	0.841	0.858	0.875
			0.8	0.900	0.870	0.823	0.823	0.840	0.857
		0.5	0.2	0.972	0.969	0.902	0.895	0.907	0.916
			0.5	0.951	0.948	0.883	0.877	0.889	0.897
			0.8	0.930	0.928	0.864	0.858	0.870	0.879
		1	0.2	0.983	0.990	0.917	0.908	0.918	0.924
			0.5	0.962	0.969	0.898	0.889	0.900	0.905
			0.8	0.941	0.948	0.879	0.871	0.881	0.887
	4	0	0.2	0.947	0.922	0.871	0.871	0.889	0.906
			0.5	0.927	0.902	0.853	0.853	0.870	0.888
			0.8	0.907	0.882	0.835	0.835	0.852	0.869
		0.5	0.2	0.979	0.983	0.915	0.908	0.921	0.929
			0.5	0.958	0.962	0.896	0.890	0.902	0.910
			0.8	0.937	0.941	0.877	0.871	0.883	0.892
		1	0.2	0.990	1.000	0.930	0.921	0.932	0.937
			0.5	0.969	0.983	0.911	0.902	0.913	0.918
			0.8	0.948	0.962	0.892	0.883	0.894	0.900

(Continued)

TABLE 3.2 (CONTINUED) The Long-Term Load Influence Coefficient kcr of Circular CFDST Member

Hollow Section Ratio ψ	Nominal Confinement Factor ξ	Eccentricity Ratio e/r	Long-Term Load Ratio n	Slenderness Ratio λ					
				20	40	80	100	120	160
0.5	1	0	0.2	0.858	0.806	0.762	0.762	0.777	0.792
			0.5	0.840	0.789	0.746	0.746	0.761	0.776
			0.8	0.821	0.772	0.730	0.731	0.745	0.761
		0.5	0.2	0.886	0.860	0.800	0.795	0.805	0.813
			0.5	0.868	0.841	0.784	0.778	0.789	0.796
			0.8	0.849	0.823	0.767	0.762	0.772	0.780
		1	0.2	0.896	0.879	0.814	0.806	0.815	0.820
			0.5	0.877	0.860	0.797	0.789	0.798	0.803
			0.8	0.858	0.841	0.780	0.773	0.782	0.787
	2	0	0.2	0.873	0.835	0.789	0.789	0.805	0.820
			0.5	0.854	0.817	0.772	0.773	0.788	0.804
			0.8	0.836	0.799	0.756	0.756	0.772	0.787
		0.5	0.2	0.902	0.890	0.829	0.823	0.834	0.841
			0.5	0.883	0.871	0.811	0.806	0.817	0.825
			0.8	0.863	0.852	0.794	0.789	0.799	0.808
		1	0.2	0.912	0.910	0.843	0.834	0.844	0.849
			0.5	0.893	0.890	0.825	0.817	0.826	0.832
			0.8	0.873	0.871	0.807	0.780	0.809	0.815

(Continued)

TABLE 3.2 (CONTINUED) The Long-Term Load Influence Coefficient kcr of Circular CFDST Member

Hollow Section Ratio ψ	Nominal Confinement Factor ξ	Eccentricity Ratio e/r	Long-Term Load Ratio n	Slenderness Ratio λ					
				20	40	80	100	120	160
3		0	0.2	0.882	0.852	0.805	0.805	0.821	0.837
			0.5	0.863	0.833	0.788	0.788	0.804	0.820
			0.8	0.844	0.815	0.771	0.772	0.787	0.803
		0.5	0.2	0.911	0.908	0.846	0.839	0.851	0.859
			0.5	0.892	0.889	0.828	0.822	0.833	0.841
			0.8	0.872	0.870	0.810	0.805	0.816	0.824
		1	0.2	0.921	0.928	0.860	0.851	0.861	0.866
			0.5	0.902	0.908	0.842	0.834	0.843	0.849
			0.8	0.882	0.889	0.824	0.816	0.826	0.831
	4	0	0.2	0.888	0.864	0.817	0.817	0.833	0.849
			0.5	0.869	0.846	0.800	0.800	0.816	0.832
			0.8	0.850	0.827	0.783	0.783	0.799	0.815
		0.5	0.2	0.918	0.922	0.858	0.852	0.863	0.871
			0.5	0.898	0.902	0.840	0.834	0.845	0.854
			0.8	0.879	0.882	0.822	0.816	0.828	0.836
		1	0.2	0.928	0.942	0.872	0.864	0.873	0.879
			0.5	0.908	0.922	0.854	0.846	0.856	0.861
			0.8	0.889	0.902	0.836	0.828	0.838	0.843

(Continued)

TABLE 3.2 (CONTINUED)   The Long-Term Load Influence Coefficient k_{cr} of Circular CFDST Member

Hollow Section Ratio ψ	Nominal Confinement Factor ξ	Eccentricity Ratio e/r	Long-Term Load Ratio n	Slenderness Ratio λ					
				20	40	80	100	120	160
0.75	1	0	0.2	0.915	0.860	0.813	0.813	0.829	0.845
			0.5	0.896	0.841	0.796	0.796	0.812	0.828
			0.8	0.876	0.823	0.779	0.779	0.795	0.811
		0.5	0.2	0.946	0.917	0.854	0.848	0.859	0.867
			0.5	0.925	0.898	0.836	0.830	0.841	0.850
			0.8	0.905	0.878	0.818	0.812	0.824	0.832
		1	0.2	0.956	0.937	0.868	0.860	0.869	0.875
			0.5	0.936	0.917	0.850	0.842	0.852	0.857
			0.8	0.915	0.897	0.832	0.824	0.834	0.839
	2	0	0.2	0.931	0.890	0.842	0.841	0.858	0.875
			0.5	0.911	0.871	0.824	0.824	0.841	0.857
			0.8	0.891	0.852	0.806	0.807	0.823	0.840
		0.5	0.2	0.962	0.949	0.884	0.877	0.889	0.897
			0.5	0.942	0.929	0.865	0.859	0.871	0.879
			0.8	0.921	0.909	0.847	0.841	0.853	0.861
		1	0.2	0.973	0.970	0.899	0.890	0.900	0.905
			0.5	0.952	0.950	0.880	0.872	0.882	0.887
			0.8	0.931	0.929	0.861	0.853	0.863	0.869

(Continued)

TABLE 3.2 (CONTINUED) The Long-Term Load Influence Coefficient kcr of Circular CFDST Member

Hollow Section Ratio ψ	Nominal Confinement Factor ξ	Eccentricity Ratio e/r	Long-Term Load Ratio n	Slenderness Ratio λ					
				20	40	80	100	120	160
	3	0	0.2	0.940	0.908	0.859	0.859	0.876	0.893
			0.5	0.920	0.889	0.841	0.841	0.858	0.875
			0.8	0.900	0.870	0.823	0.823	0.840	0.857
		0.5	0.2	0.972	0.969	0.902	0.895	0.907	0.916
			0.5	0.951	0.948	0.883	0.877	0.889	0.897
			0.8	0.930	0.928	0.864	0.858	0.870	0.879
		1	0.2	0.983	0.990	0.917	0.908	0.918	0.924
			0.5	0.962	0.969	0.898	0.889	0.900	0.905
			0.8	0.941	0.948	0.879	0.871	0.881	0.887
	4	0	0.2	0.947	0.922	0.871	0.871	0.889	0.906
			0.5	0.927	0.902	0.853	0.853	0.870	0.888
			0.8	0.907	0.882	0.835	0.835	0.852	0.869
		0.5	0.2	0.979	0.983	0.915	0.908	0.921	0.929
			0.5	0.958	0.962	0.896	0.890	0.902	0.910
			0.8	0.937	0.941	0.877	0.871	0.883	0.892
		1	0.2	0.990	1.004	0.930	0.921	0.932	0.937
			0.5	0.969	0.983	0.911	0.902	0.913	0.918
			0.8	0.948	0.962	0.892	0.883	0.894	0.900

a The intermediate values can be calculated by linear interpolation; for the members with slenderness ratio under 20, the influence coefficient can be treated as the case with slenderness ratio equals to 20.

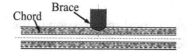
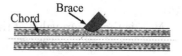

FIGURE 3.1 Typical connection area of CFDST members under local bearing.

3.6.2 For CFDST members which may have a local bearing failure in the joint area, the local bearing capacity should be checked. As shown in Figure 3.2, the bearing capacity should meet the following requirements:

$$N_{LF} \leq N_{uLF} \qquad (3.28)$$

$$N_{uLF} = (1-\psi)\beta_c\beta_1 f_c \frac{A_1}{\sin\theta} \qquad (3.29)$$

$$\beta_1 = \sqrt{A_b / A_1} \qquad (3.30)$$

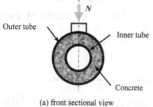

(a) front sectional view

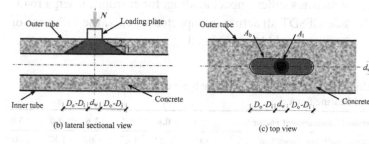

(b) lateral sectional view   (c) top view

FIGURE 3.2 The diagram for the load transfer of local bearing force in the connection area of CFDST.

where:

N_{LF} is the design value of the lateral local compression in the joint area of CFDST members

N_{uLF} is the design value of bearing capacity for CFDST members to resist lateral local compression

f_c is the design value of the concrete strength

θ is the angle between the brace and the chord

β_1 is the strengthen coefficient for the concrete subjected to lateral local compression

A_1 is the area of local compression for CFDST (for hollow branch pipe, the area is the same value as for the solid branch pipe which has identical diameter), $A_1 = \dfrac{\pi d_w^2}{4}$

d_w is diameter of the brace which transfers lateral forces

A_b is the maximum design distribution area for lateral partial compression, $A_b = \dfrac{A_1}{\sin\theta} + 2d_w(D_o - D_i)$

β_c is the influence coefficient for concrete strength subjected to lateral local compression (refer to Table 3.3), when calculating the nominal confinement factor ξ for CFDST members, $A_{so} = A_1$; $A_{ce} = A_1\sqrt{A_b / A_1}$.

3.7 IMPACT

3.7.1 To avoid the potential damage or failure when CFDST structures suffer impact loading, for example, when a roadside CFSDT structure is impacted by cars, the influence of impact should be considered.

TABLE 3.3 The Influence Coefficient β_c for Concrete Strength Subjected to Local Bearing Forces[a]

Nominal Confinement Factor ξ	0.4	0.8	1.2	1.6	2.0	≥ 3.0
The strength coefficient of concrete β_c	1.07	1.22	1.47	1.67	1.87	2.00

[a] The intermediate values can be calculated by linear interpolation.

3.7.2 When impact was caused by cars with speeds under 30 m/s, the bending capacity of the CFDST column should meet the requirements of provision 3.2.

3.7.3 The impact resistance capacity of CFDST members drops sharply when the hollow ratio reaches 0.8. It is recommended that the hollow ratio be less than 0.7 for a CFDST in risk of impact in practice.

3.8 EFFECTS OF CORROSION

3.8.1 During the design of CFDST structures, the material, construction measures and anti-corrosion measures should be reasonably selected to meet the corresponding requirements.

3.8.2 When the outer tube of CFDST is corroded in a corrosive environment, the bearing capacity after corrosion can be checked approximately using the "equivalent section method." The specific method uses the equivalent parameters after corrosion to replace the original parameters during the calculation of bearing capacity. The formulae for the calculation of the equivalent parameters are as follows:

$$t_{oe} = t_o - \Delta t_o \tag{3.31}$$

$$D_e = D_o - 2\Delta t_o \tag{3.32}$$

$$\alpha_{ne} = \frac{A_{soe}}{A_{ce}} \tag{3.33}$$

$$\psi_e = \frac{D_i}{D_e - 2t_{oe}} \tag{3.34}$$

$$\xi_e = \frac{A_{soe} \cdot f_{y,\text{outer}}}{A_{ce} \cdot f_{ck}} = \alpha_{ne} \cdot \frac{f_{y,\text{outer}}}{f_{ck}} \tag{3.35}$$

where:

Δt_o is the thickness loss of the outer tube under corrosion

t_{oe} is the equivalent wall thickness of the outer tube

D_e is the equivalent external diameter of the outer tube

A_{soe} is the equivalent sectional area of the outer tube

α_{ne} is the equivalent nominal steel ratio

ψ_e is the equivalent hollow section ratio

ξ_e is the equivalent standard value of the nominal confinement factor

3.8.3 For localized corrosion on the outer tube of CFDST members, such as pitting and point corrosion, the "equivalent section method" could be adopted with the largest thickness loss along the cross section taken as Δt_o in 3.8.2.

3.9 LIMITATION AND CALCULATION OF CONCRETE IMPERFECTION

3.9.1 Separation may occur between the concrete and the tube in CFDST members during construction and service, which would influence the composite action between the concrete and the steel and thus decrease the bearing capacity. The two most common types of separations are the circumferential gap (shown in Figure 3.3a) and the circular-segment gap (shown in Figure 3.3b).

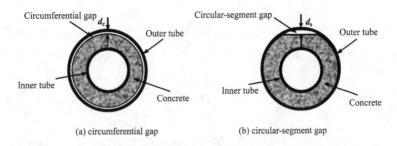

(a) circumferential gap   (b) circular-segment gap

FIGURE 3.3   Separation between the concrete and the tube in CFDST.

3.9.2 The ratio of the circumferential gap and the circular-segment gap should be calculated from Equations (3.36) and (3.37), respectively.

$$\psi_c = \frac{2d_c}{D_o} \tag{3.36}$$

$$\psi_s = \frac{d_s}{D_o} \tag{3.37}$$

where:

ψ_c is the ratio of the circumferential gap
d_c is the distance between the concrete and the tube along the radial direction, as shown in Figure 3.3a
ψ_s is the ratio of the circular-segment gap
d_s is the distance between the concrete and the tube along the radial direction, as shown in Figure 3.3b

3.9.3 The ratio of the circumferential gap should not exceed 0.05%; if it does, the member is not qualified for confinement. The possible composite effect should be ignored when this ratio exceeds 0.05% and the member should be considered as a steel member.

3.9.4 When the ratio of the circular-segment gap is less than 2.4%, the strength reduction of the member can be ignored. When the ratio exceeds 2.4%, the calculation of bearing capacity should be conducted according to the equations below:

$$N \le k \cdot N_u \tag{3.38}$$

$$k = \begin{cases} 1 - \psi(1.42\xi + 0.44) & \xi \le 1.24 \\ 1 - \psi(4.66 - 1.97\xi) & \xi > 1.24 \end{cases} \tag{3.39}$$

where:

N_u is the axial compressive bearing capacity of CFDST with no separation, calculated by Equation (3.7) in 3.2.

CHAPTER 4

Joints and Connections

4.1 GENERAL DESCRIPTION

4.1.1 Joint and connection design of CFDST structures should meet the requirements of strength, stiffness, stability and seismic performance to ensure that the force can be transferred effectively and that the inner tube, the outer tube and the concrete could work together, making it easier for manufacturing and installation. Proper concrete placement methods should be chosen to ensure the density of concrete in the joint area.

4.1.2 The construction of the steel structure should be simple and induce a clear load transfer path. The center lines (or bolt alignments) of each stressed bar should converge on one point so as to reduce eccentricity. Connections of the steel tubes should use a flange plate or axis-rib configuration. The inner tube and the outer tube can be connected with the partition plate bolt.

4.1.3 The weld check and the tensile capacity check in the connection between branches of the built-up members in CFDST structures should be conducted according to the regulations in EN 1993-1-8 (Chapter 4).

4.2 JOINTS

4.2.1 Diagonal web member configuration (Figure 4.1) should be used for lattice columns under eccentric pressure, where CFDST members are used as chord components and hollow

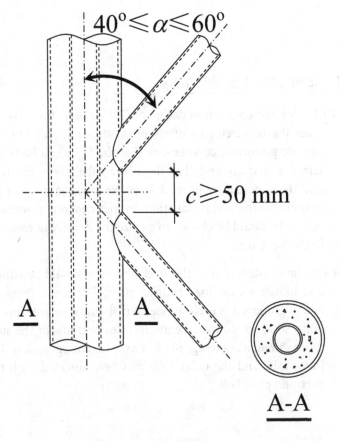

FIGURE 4.1   Diagonal web member joints.

steel tubes as web members. Transverse web member form (Figure 4.2) should be used when the spacing of column components is not enough for fabrication of diagonal web member configuration, while the mixed form of the diagonal and transverse web member (Figure 4.3) should be used when the spacing of column components is sufficient. The construction of the lattice column should be in accordance with the following provisions:

1. Lattice column with diagonal web member configuration:

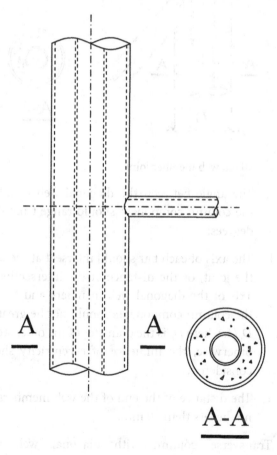

A   A   A-A

FIGURE 4.2   Transverse web member joints.

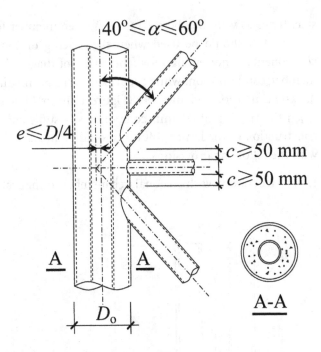

FIGURE 4.3   Mixed web member joints.

a. The angle between the diagonal web member and the column component should range from 40 to 60 degrees.

b. The axis of each bar should intersect at the center of the joint, or the distance of the intersection of the axis of the diagonal web members and the axis of the column components should not be greater than $D_0/4$, where D_0 is the diameter of the outer steel tube; otherwise, the influence of eccentricity should be considered;

c. The distance of the end of the web members should not be less than 50 mm.

2. Transverse column with diagonal web member configuration:

a. The distance between the center of adjacent web members should not be greater than 4 times the distance between the center of the adjacent column components;

b. The cross-sectional area of the hollow steel tube web member should not be less than a quarter of that of the outer steel tube of the column component;

c. The slenderness ratio of the web members should not be larger than half of that of a single column component.

3. Hollow steel tubes can be used as web members.

4. The web members and column components should be welded directly, and holes should not be drilled in the steel surface of the column components.

4.2.2 Diaphragms should be added at the positions carrying a horizontal load and at the ends of the transport units for three and four components lattice column. The distance between the diaphragms should not be greater than the smaller of 9 times the sectional diameter of the largest column or 8 m.

4.2.3 The rigid beam-column joint in a frame structure using CFDST columns should adopt the configuration of strengthening ring plates. When the outer diameter of the steel tube is large enough for installation of inner baffle plate, the form of the inner baffle plate can be used, but the pouring of concrete between the inner and outer tubes should be ensured not to be affected.

When the beam is an I-section steel beam or a steel-concrete composite beam, the construction of the joint is shown in Figure 4.4. Design of the joint should be in accordance with EN 1993-1-8.

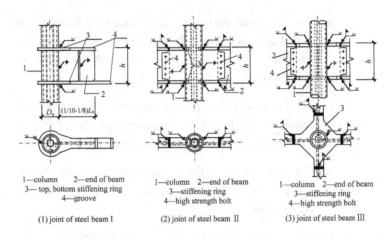

1—column   2—end of beam
3— top, bottom stiffening ring
4—groove

(1) joint of steel beam I

1—column   2—end of beam
3—stiffening ring
4—high strength bolt

(2) joint of steel beam II

1—column   2—end of beam
3—stiffening ring
4—high strength bolt

(3) joint of steel beam III

FIGURE 4.4   Joint of steel beam.

4.2.4 The rigid joints between a circular section CFDST structure and the stiffening ring plate generally includes four types of configurations, shown in Figure 4.5.

When the stiffening ring plate is bearing tension N in the direction of the beam, N can be calculated from the following formula:

$$N = \frac{M}{h} + N_b \tag{4.1}$$

$$M = M_c - \frac{V \cdot D_o}{3}, \quad M \geq 0.7 M_c \tag{4.2}$$

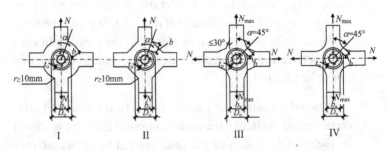

FIGURE 4.5   Types of the stiffening ring plate of circle in circle CFDST.

where:

- M is the design value of the bending moment at the end of the beam

- N_b is the tension of the ring plate caused by the axial force of the beam

- h is the height of the section of the end of the beam

- M_c is the design value of the bending moment of the support of the beam at the center line of the column

- V is the design value of the shear force of the end of the beam at the axis of the column corresponding to the M_c

- D_o is the outer diameter of the outer circular steel tube

1. Calculation of width (b_s) and thickness (t_1) of the stiffening ring plate

    a. The width of the top ring plate connecting RC beam (b_s) should be 20 to 40 mm less than the width of the beam; the width of the bottom ring plate should be 20 to 40 mm greater than the width of the beam; the width of the ring plate connecting steel beam (b_s) should be equal to the width of the beam flange

    b. The thickness of the steel plate connecting the reinforced concrete beam (t_1) can be calculated from the following formula, and strength of the weld should be checked

$$t_1' = \frac{A_s \cdot f}{b_s \cdot f_1} \tag{4.3}$$

where:

- A_s is the area of the steel bar in the section with negative bending moment at the end of the beam

- f is the design value of tensile strength of the steel bar in the section with negative bending moment at the end of the beam

f_1 is the design value of tensile strength of the steel of the stiffening ring plate

The thickness of the ring plate connecting the steel beam (t_1) should be determined according to the axial tension of the beam flange.

2. Calculation of the width of the control section of the stiffening ring plate (b)

   a. For I type and II type stiffening ring plate, b should satisfy the following equations:

$$b \geq F_1 \cdot \frac{N}{t_1 \cdot f_1} - F_2 \cdot b_e \cdot \frac{t_o \cdot f_{outer}}{t_1 \cdot f_1} \qquad (4.4)$$

$$F_1 = \frac{0.93}{\sqrt{2 \sin^2 \alpha + 1}} \qquad (4.5)$$

$$F_2 = \frac{1.74 \sin \alpha}{\sqrt{2 \sin^2 \alpha + 1}} \qquad (4.6)$$

$$b_e = (0.63 + 0.88 \frac{b_s}{D_o}) \cdot \sqrt{D_o \cdot t_o} + t_1 \qquad (4.7)$$

where:

α is the angle between the direction of tension and the calculation cross section

b_e is the effective width of the outer tube wall of the column participating in the work of the stiffening ring (Figure 4.6)

t_o is the thickness of the outer tube of the column

f_o is the design value of the steel strength of the outer tube of the column

F_1, F_1 are geometrical coefficients

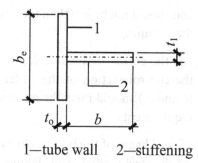

1—tube wall   2—stiffening

FIGURE 4.6   Effective width of the outer tube wall of the circular column.

b. For the III type and IV type stiffening ring plate, b should satisfy the following formula:

$$b \geq (1.44 + \beta_h) \cdot \frac{0.392 N_{x,max}}{t_1 \cdot f_1} - 0.864 b_e \cdot \frac{t_0 \cdot f_{outer}}{t_1 \cdot f_1} \qquad (4.8)$$

$$\beta_h = \frac{N_y}{N_{x,max}} \leq 1 \qquad (4.9)$$

where:

β_h is the ratio of the vertical biaxial tension of the stiffening ring plate when the stiffening ring plate is bearing uniaxial tension, $\beta_h = 0$

$N_{x,max}$ is the tension caused by the combination of the most adverse effects in the X direction

N_y is the tension in the Y direction working with $N_{x,max}$ at the same time

c. Construction of the stiffening ring plate:

    i. $0.25 \leq b_s/D_0 \leq 0.75$.

    ii. $0.1 \leq b_s/D_0 \leq 0.35$, $b_s/t_1 \leq 10$. The length of the column in the connection area is calculated from the top and bottom stiffening ring plate and should not be less than the outer diameter of the column.

iii. For the corner and side columns, the width of the control section of the stiffening ring plate (b and h_s) should meet the minimum structural requirements.

4.2.5 The web of steel beam of the beam-column joint and the shear stress of the steel Corbel plate (Figure 4.7) should be calculated as follows:

$$\tau = 0.6 \frac{V_{max}}{h_j \cdot t_o} \cdot \lg\left(\frac{2r_c}{b_j}\right) \le f_{jv} \qquad (4.10)$$

$$b_j = t_w + 2h_f \qquad (4.11)$$

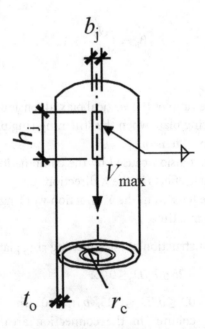

FIGURE 4.7 The calculation diagram of the tube stress of CFDST.

where:

V_{max} is the maximum shear force that the web of end of the beam or a Corbel plate can bear

h_j is the height of the web or the plate

r_c is the size of the nominal core concrete section, and $r_c = D_o/2 - t_o$, t_o is the thickness of the outer tube

f_{jv} is the design value of the shear strength of the weld

b_j is the width of the fillet weld

t_w is the thickness of the weld (plate)

h_f is the leg size of the fillet weld

4.2.6 Parts such as bracket and hanger can be welded on the outer tube of CFDST. When supporting heavy brackets and hangers on the tube wall, the stiffening ring configuration should be adopted, the strength of the outer tube should be checked and the influence of the additional load of the hanger on the CFDST member should be considered as well.

4.2.7 The joint design for earthquake regions should meet the following requirements:

1. Only III~VI type stiffening ring beam-column joints with inner diaphragm in 4.2.4 should be used.

2. When using a joint of steel beam, the section design of the beam should meet the relevant rules of the standards EN 1993-1-8 and EN 1998-1.

3. The seismic resistance checking of the stiffening ring plate can be conducted according to the relevant rules for steel structure in EN 1993-1-8.

4. Joints should meet the following requirements:

   a. Smooth shape curve with no cracks or nicks should be ensured for the manufacture of the stiffening ring plate;

b. The horizontal weld with the same strength for the base metal should be used between tubes of the joint part and the column tube;

c. Groove welding should be used for the butt welding between the stiffening ring plate and the steel beam flange.

5. For the maximum stress area that may produce plastic hinges, welding should be avoided.

4.2.8 The rigid joints of CFDST beams and columns can adopt the following configurations (Figures 4.8 through 4.10). Blind bolts could also be used for the beam-column connections,

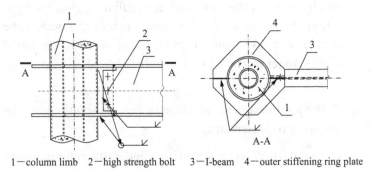

1—column limb   2—high strength bolt   3—I-beam   4—outer stiffening ring plate

FIGURE 4.8   Joint of outer stiffening ring plate form.

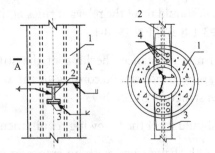

1—column limb   2—steel beam   3—flange reinforcing plate   4—hole on top, bottom flange

FIGURE 4.9   Joint through half core of Cantilever beam form.

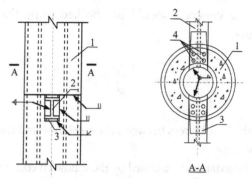

1—column   2—steel beam   3—flange reinforcing plate   4—hole on top, bottom flange

FIGURE 4.10   Joint of drop form.

and the steel beams can be fastened by blind bolts to either the inner or outer skin of the CFDST columns.

4.2.9 The whole or part of the shear force of the beam end in the joint area is transferred to the inner tube and the concrete in the form of shear stress in CFDST, as shown in Figure 4.11. In the frame, the bond stress between the inner tube, outer

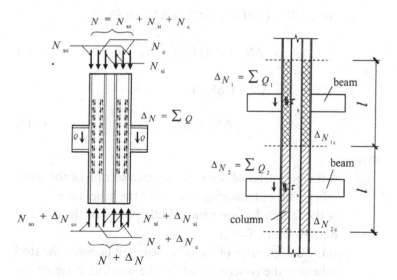

FIGURE 4.11   Schematic diagram of stress transfer.

tube and concrete should be checked using the following formula:

$$\tau_s = \frac{\Delta N_{ic}}{l_c \cdot l} \le f_b \tag{4.12}$$

where:

τ_s is the bond stress between the steel tube and the sandwich concrete

ΔN_{ic} is the axial load sustained by the sandwich concrete transmitted from the beam of floor i which is connected to the column

The calculation method of ΔN_{ic} is as follows:

Assuming that the resultant force of the shear force at the beam end is ΔN_1, and N_1 is the axial force acting on the column:

a. When $N_1 \ge 0.85 f_c \cdot A_c$, the shear force at the beam end is carried by the steel tube, and it is unnecessary to check the bond strength;

b. When $N_1 < 0.85 f_c \cdot A_c$, but $N_1 + \Delta N_1 > 0.85 f_c \cdot A_c$,

$$\Delta N_{ic} = 0.85 f_c \cdot A_c - N_1 \tag{4.13}$$

c. When $N_1 + \Delta N_1 < 0.85 f_c \cdot A_c$,

$$\Delta N_{ic} = \Delta N_1 \tag{4.14}$$

where:

l_c is the perimeter of the inner surface section of the outer tube or the outer surface section of the inner tube

l is the length between the middle points of the top and bottom floor column

f_b is the design value of the bond strength between the steel tube and the concrete, and for the bonding between the circular steel tube and the concrete, $f_b = 0.225$ N/mm^2

When the bond between the steel tube and the concrete does not meet the requirements, appropriate construction measures should be taken. For instance, shear connectors should be provided between the steel and concrete interfaces.

4.3 FLANGE CONNECTIONS

The following configuration (Figure 4.12) could be adopted in the flange connection of CFDST structures. The flange bolts, flange plate and weld in the flange connections can be designed according to EN 1998-6 and BS EN 1092. Other flange connections could also be used as construction conditions allow. It is recommended that flange connection to the inner and outer tubes should not occur in the same section; a minimum lap distance of 300 mm is recommended.

4.4 PLATE CONNECTIONS

Typical connections between vertical structural components and horizontal components as well as the connections between main components and inclined components in CFDST structures, are shown in Figure 4.13. These can be optimized by designers. Designs should be conducted according to EN 1993-1-8 in terms of the strength of the connection plate, stability, the weld joint and the bolt.

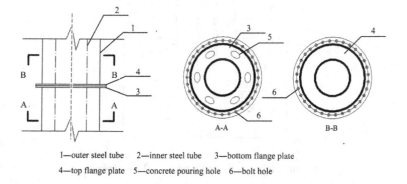

1—outer steel tube   2—inner steel tube   3—bottom flange plate
4—top flange plate   5—concrete pouring hole   6—bolt hole

FIGURE 4.12   Schematic diagram of outer flange connections.

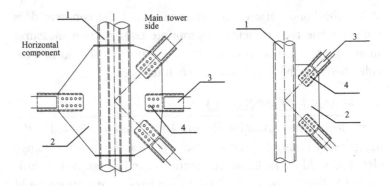

FIGURE 4.13   Schematic diagram of plate connections.

4.5 COLUMN-FOOT CONNECTIONS

4.5.1 Typical column bases for CFDST include pinned column bases and fixed column bases.

a. The design of pinned column bases should be according to EN 1993-1-8 (Chapter 6).

b. The rigid column bases include plug-in column bases and embedded column bases.

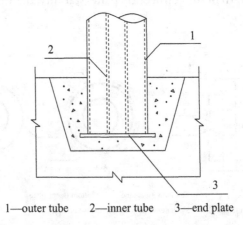

1—outer tube   2—inner tube   3—end plate

FIGURE 4.14   Schematic diagram of plug-in column base.

a. A plug-in column base is shown in Figure 4.14. The design of the base cup is the same as that for reinforced concrete (RC) structures and should meet the construction demands.

b. The design of an embedded column base with steel plate should be undertaken according to EN 1993-1-8.

The column should be embedded by concrete with thickness no less than 50 mm. The top of the column should not be less than 200 mm higher than the ground. Studs or stub rebars can be welded to the outer surface of the outer tube to increase the bond between the composite CFDST columns and the outer concrete.

CHAPTER 5

Fire-Resistance Design

5.1 When the load ratio (n) of a CFDST column does not exceed the strength index (k_t), no extra measures are needed for the fire-resistance design, where the load ratio (n) was defined as the ratio of the axial compressive load applied in the fire situation to the load-bearing capacity at ambient temperature. The strength indexes for CFDST columns subjected to different fire-resistance ratings can be obtained from Table 5.1.

5.2 For CFDST columns which do not meet provision 5.1, measures such as spraying fireproofing coatings or other effective outer cover prevention should be employed according to the fire-resistance design method, so that the fire-resistant level as well as the fire resistance could meet the requirements of corresponding design guideline EN 1994-1-2.

When the fire-proof coating with properties given in Table 5.2 is used for a fire-proof protection layer, the

TABLE 5.1 The Strength Index k_t for Circular CFDST Column

λ	D_o (mm)	ψ = 0.25 Fire Resistance (h)						ψ = 0.5 Fire Resistance (h)						ψ = 0.75 Fire Resistance (h)					
		0.5	1.0	1.5	2.0	2.5	3.0	0.5	1.0	1.5	2.0	2.5	3.0	0.5	1.0	1.5	2.0	2.5	3.0
10	200	0.59	0.48	0.44	0.40	0.37	0.34	0.56	0.43	0.38	0.33	0.29	0.25	0.53	0.37	0.30	0.23	0.19	0.15
	400	0.61	0.51	0.48	0.45	0.42	0.39	0.58	0.46	0.41	0.36	0.33	0.29	0.54	0.40	0.32	0.26	0.21	0.17
	600	0.63	0.54	0.51	0.48	0.46	0.44	0.60	0.48	0.43	0.39	0.35	0.32	0.56	0.41	0.34	0.28	0.23	0.19
	800	0.65	0.55	0.53	0.50	0.48	0.47	0.62	0.50	0.45	0.41	0.38	0.35	0.58	0.43	0.35	0.29	0.24	0.21
	1000	0.68	0.56	0.54	0.52	0.50	0.49	0.64	0.51	0.46	0.42	0.39	0.36	0.60	0.44	0.36	0.30	0.25	0.21
	1200	0.70	0.57	0.55	0.53	0.51	0.50	0.66	0.52	0.47	0.43	0.40	0.37	0.62	0.44	0.37	0.31	0.26	0.22
	1400	0.73	0.58	0.56	0.54	0.52	0.50	0.69	0.52	0.47	0.44	0.40	0.38	0.64	0.45	0.37	0.31	0.26	0.22
20	200	0.58	0.35	0.29	0.24	0.19	0.15	0.55	0.31	0.25	0.20	0.15	0.11	0.51	0.27	0.20	0.14	0.10	0.07
	400	0.59	0.40	0.36	0.32	0.28	0.24	0.56	0.36	0.31	0.26	0.22	0.18	0.53	0.31	0.24	0.19	0.14	0.11
	600	0.61	0.44	0.40	0.37	0.34	0.31	0.58	0.39	0.34	0.30	0.26	0.23	0.54	0.34	0.27	0.22	0.17	0.14
	800	0.64	0.46	0.43	0.41	0.38	0.36	0.60	0.41	0.37	0.33	0.30	0.27	0.56	0.36	0.29	0.24	0.19	0.16
	1000	0.66	0.48	0.45	0.43	0.41	0.39	0.63	0.43	0.39	0.35	0.32	0.29	0.58	0.37	0.30	0.25	0.21	0.17
	1200	0.68	0.49	0.47	0.45	0.43	0.41	0.65	0.44	0.40	0.36	0.33	0.31	0.60	0.38	0.31	0.26	0.22	0.18
	1400	0.71	0.50	0.48	0.46	0.44	0.43	0.68	0.45	0.41	0.37	0.34	0.32	0.63	0.38	0.32	0.27	0.22	0.19
40	200	0.43	0.23	0.14	0.06	0.00	0.00	0.40	0.21	0.12	0.05	0.00	0.00	0.38	0.18	0.10	0.03	0.00	0.00
	400	0.47	0.30	0.24	0.17	0.11	0.06	0.45	0.27	0.20	0.14	0.09	0.04	0.42	0.23	0.16	0.10	0.06	0.03
	600	0.50	0.35	0.30	0.25	0.21	0.17	0.48	0.31	0.26	0.21	0.16	0.13	0.45	0.27	0.20	0.15	0.11	0.08

(Continued)

TABLE 5.1 (CONTINUED) The Strength Index kt for Circular CFDST Column

λ	D_o (mm)	ψ = 0.25						ψ = 0.5						ψ = 0.75					
		Fire Resistance (h)						Fire Resistance (h)						Fire Resistance (h)					
		0.5	1.0	1.5	2.0	2.5	3.0	0.5	1.0	1.5	2.0	2.5	3.0	0.5	1.0	1.5	2.0	2.5	3.0
	800	0.53	0.38	0.34	0.31	0.28	0.25	0.51	0.34	0.29	0.25	0.22	0.19	0.47	0.29	0.23	0.18	0.14	0.11
	1000	0.56	0.40	0.38	0.35	0.33	0.30	0.53	0.36	0.32	0.28	0.25	0.23	0.49	0.31	0.25	0.20	0.16	0.13
	1200	0.59	0.42	0.40	0.38	0.36	0.34	0.56	0.38	0.34	0.30	0.28	0.25	0.52	0.32	0.27	0.22	0.18	0.15
	1400	0.62	0.43	0.41	0.39	0.37	0.36	0.59	0.39	0.35	0.32	0.29	0.27	0.55	0.33	0.28	0.23	0.19	0.16
60	200	0.29	0.16	0.04	0.00	0.00	0.00	0.28	0.15	0.03	0.00	0.00	0.00	0.26	0.13	0.03	0.00	0.00	0.00
	400	0.35	0.25	0.16	0.08	0.00	0.00	0.33	0.23	0.14	0.06	0.00	0.00	0.31	0.19	0.11	0.05	0.00	0.00
	600	0.38	0.31	0.25	0.19	0.13	0.08	0.36	0.28	0.21	0.15	0.10	0.06	0.34	0.24	0.17	0.11	0.07	0.03
	800	0.40	0.35	0.31	0.27	0.23	0.19	0.38	0.32	0.26	0.22	0.17	0.14	0.36	0.27	0.21	0.15	0.11	0.08
	1000	0.43	0.38	0.35	0.32	0.29	0.26	0.41	0.35	0.30	0.26	0.22	0.20	0.38	0.30	0.24	0.19	0.15	0.12
	1200	0.46	0.41	0.38	0.35	0.33	0.31	0.43	0.36	0.32	0.29	0.26	0.23	0.40	0.31	0.25	0.21	0.17	0.14
	1400	0.49	0.42	0.40	0.37	0.36	0.34	0.46	0.38	0.34	0.30	0.28	0.25	0.43	0.32	0.27	0.22	0.18	0.15

PS: the intermediate values can be obtained by linear interpolation.

TABLE 5.2 The Properties of Fire-Proof Coating

			Requirement			
Bond strength (MPa)			≥0.04			
Compressive strength (MPa)			≥0.3			
Dry density (kg/m³)			≤500			
Thermal conductivity [W/(m·K)]			≤0.1160			
Water resistance (h)			≥24			
Cyclic freeze-thaw resistance (time)			≥15			
Fire resistance	Thickness of fire-proof protection layer (mm)	15	20	30	40	50
	Fire-resistant time not lower than (h)	1.0	1.5	2.0	2.5	3.0

thickness of the fire-proof protection layer (a) for CFDST columns should be calculated as follows:

$$a = k_{h1} \cdot a_1 \geq 7\,\text{mm} \tag{5.1}$$

where k_{h1} is the influence coefficient of the hollow ratio, which can be calculated from Equation (5.2).

$$k_{h1} = \exp[(\alpha \cdot \lambda_o^2 + \beta \cdot \lambda_o + \gamma) \cdot \psi^2] \tag{5.2}$$

where $\lambda_o = \lambda / 40$.

$$\alpha = 0.05t^2 - 0.24t - 0.17; \quad \beta = -0.12t^2 + 0.62t + 0.43; \quad \gamma = 0.208t + 0.824$$

The factor a_1 in Equation (5.1) may be calculated as follows:

For CFDST members with CHS outer and CHS inner:

$$a_1 = k_{LR} \cdot (19.2t + 9.6) \cdot C^{-(0.28 - 0.0019\lambda)} \tag{5.3}$$

where:

When $k_t < 0.77$ and $k_t < n < 0.77$, $k_{LR} = p \cdot n + q$

When $k_t < 0.77$ and $n \geq 0.77$, $k_{LR} = 1/(3.695 - 3.5n)$

When $k_t \geq 0.77$, $k_{LR} = \omega \cdot (n - k_t)/(1 - k_t)$

where:

$p = 1/(0.77 - k_t)$; $q = k_t/(k_t - 0.77)$; $\omega = 7.2t$;

In Equations (5.1), (5.2) and (5.3):

t is the fire resistance, in h

n is the ratio of the axial compressive load applied in the fire situation to the load-bearing capacity at ambient temperature

C is the perimeter of the outer section for the CFDST member. $C = \pi D_o$, where $C = 628 \sim 6280$ mm, which means $D_o = 200 \sim 2000$ mm

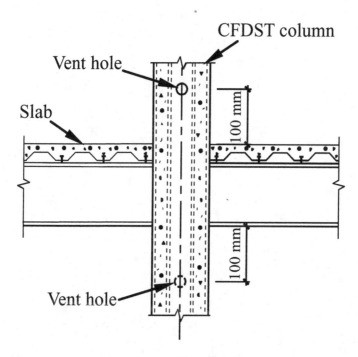

FIGURE 5.1   Schematic diagram of exhaust vents position.

λ is the slenderness ratio, with the range of 10~80

ψ is the hollow ratio, with the range of 0~0.75

k_t is the strength index for the CFDST column. Refer to Table 5.1

5.3 To ensure the release of moisture in the inner concrete during fire, a vent hole with the diameter of 10 mm should be set on the outer tube of columns in each story, the positions of which are suggested to be above and below the intersection of the columns and slabs, and the vent holes are suggested to be set central and symmetrical, as shown in Figure 5.1. Two centrally symmetrical vent holes should be set on each member of the CFDST truss, and a couple of vent holes should be added when the length of the member exceeds 4 m.

CHAPTER 6

Construction

6.1 GENERAL DESCRIPTION

6.1.1 Procedures for the manufacturing of steel tubes should be organized according to the detailed drawings and documents of design. As for complicated members, experiments should be conducted to ensure the reliability of construction techniques.

6.1.2 If the hollow steel tubes are installed first, followed by the pouring of the concrete, the strength and stability of the hollow steel tubes under the construction load should be checked.

6.1.3 Steel members should be fixed and checked before the pouring of sandwiched concrete.

6.1.4 Appropriate measures should be taken in accordance with EN 1993-1-8 and EN 10025-5 to prevent the corrosion of steel tubes.

6.2 FABRICATION OF TUBULAR MEMBERS

6.2.1 Construction and manufacture of steel tubes should follow EN 1993-1-3.

6.2.2 The inner surface of the steel tube should be kept clean and free from oil stain after manufacturing.

6.2.3 The ends of tubes should be sealed during installation in case debris may enter the tube. Concrete strength should reach 50% of the design value before installation in a precast CFDST component.

6.3 CONCRETE PLACEMENT

6.3.1 The quality of the sandwiched concrete in a CFDST component should be monitored during all stages of the concrete pouring.

6.3.2 The tubes should be properly fixed and checked before pouring the concrete. Dust and water in the tube should be appropriately cleared. Concrete should not be poured into the inner tube.

6.3.3 The methods of concrete placement include pump filling, vibrating tube by tube along the whole length, filling by embedded pipe, gravity filling and so on. Trials may be done to test the compactness of the concrete so that the proper method of placement is chosen.

6.3.4 Not only the strength but also the slump of the concrete should be considered for the design of the concrete mix. The water-cement ratio should follow the strength level of the concrete and be tested by experiments. It should also satisfy the workability requirements of the concrete if the pump-filling method is taken.

6.3.5 The concrete must be poured continuously. When construction gaps are necessary, the ends of the tube should be sealed to keep the tube free from water, dust and oil.

6.3.6 To avoid the bounce of aggregate during the pouring process, cement grout with the same strength of the concrete with thickness of 100–200 mm will be placed before the pouring of the concrete.

6.3.7 Concrete may spill from the tube before assembling the diaphragms with exhaust vents between stories or end plate using spot welding. Patch welding is made after the concrete reaches 50% of the designed strength value. An alternative method is to pour the concrete up to a height slightly lower than the tube end. Cement grout with the same strength of the concrete is poured to fill the tube after the concrete reaches 60% of the design strength, and welding of diaphragm or end plate is finished sequentially.

6.3.8 Supersonic wave could be used to monitor the compactness of the sandwiched concrete. Local drilling and grouting technique should be used to remedy defect.

6.4 QUALITY CHECK

Quality check of CFDST structures should follow codes and specifications including EN 1992, EN 1993, EN 1994.

References

EN 1990, 2002. Eurocode – Basis of structural design.

EN 1992-1-1, 2004. Eurocode 2: Design of concrete structures – Part 1-1: General rules and rules for buildings.

EN 1993-1-1, 2005. Eurocode 3: Design of steel structures – Part 1-1: General rules and rules for buildings.

EN 1993-1-3, 2005. Eurocode 3: Design of steel structures – Part 1-1: General rules – Supplementary rules for cold-formed members and sheeting.

EN 1993-1-8, 2005. Eurocode 3: Design of steel structures – Part 1-8: Design of joints.

EN 1994-1-1, 2004. Eurocode 4: Design of composite steel and concrete structures – Part 1: General rules and rules for buildings.

EN 1994-1-2, 2005. Eurocode 4: Design of composite steel and concrete structures - Part 1-2: General rules - Structural fire design.

EN 1998-1-1, 2004. Eurocode 8: Design of structures for earthquake resistance – Part 1: General rules, seismic actions and rules for buildings.

EN 1998-1-6, 2004. Eurocode 8: Design of structures for earthquake resistance – Part 6: Towers, masts and chimneys.

Goode CD, Lam D, 2011. Concrete-filled steel tube columns-tests compared with Eurocode 4. *Composite Construction in Steel and Concrete*, 6: 317–325.

Han LH, 2016. *Concrete Filled Steel Tubular Structures – Theory and Practice* (Third Edition). Beijing: China Science Press.

Han LH, Tao Z, Huang H, Zhao XL, 2004. Concrete-filled double skin (SHS outer and CHS inner) steel tubular beam-columns. *Thin-Walled Structures*, 42(9): 1329–1355.

Han LH, Li W, Bjorhovde R, 2014. Developments and advanced applications of concrete-filled steel tubular (CFST) structures: Members. *Journal of Constructional Steel Research*, 100: 211–228.

Hicks SJ, Newman GM, Edwards M, et al., 2002. *Design Guide for Concrete Filled Columns*. Corby, Northants: Corus Tubes.

Lam D, Williams CA, 2004. Experimental study on concrete filled square hollow sections. *Steel and Composite Structures*, 4(2): 95–112.

Morino S, Tsuda K, 2003. Design and construction of concrete-filled steel tube column system in Japan. *Earthquake Engineering and Engineering Seismology*, 4(1): 51–73.

Qu. X, Chen, Z, Nethercot, D.A., Gardner L., Theofanous, M., 2015. Push-out tests and bond strength of rectangular CFST columns. *Steel and Composite Structures*, 19(1): 21–41.

Roeder CW, 1998. Overview of hybrid and composite systems for seismic design in the United States. *Engineering Structures*, 20(4): 355–363.

Structural Stability Research Council, Task Group 20, 1979. Specification for the design of steel-concrete composite columns. *Engineering Journal*, AISC, Fourth Quarter.

Tao Z, Han LH, Zhao XL, 2004. Behaviour of concrete-filled double skin (CHS inner and CHS outer) steel tubular stub columns and beam-columns. *Journal of Constructional Steel Research*, 60(8): 1129-1158.

Yagishita F, Kitoh H, Sugimoto M, Tanihira T, Sonoda K, 2000. Double skin composite tubular columns subjected to cyclic horizontal force and constant axial force. *Proceedings of the 6th International Conference on Advances in Steel and Concrete Composite Structures*, pp. 497–503.

Yang H, Lam D, Gardner L, 2008. Testing and analysis of concrete-filled elliptical hollow sections. *Engineering Structures*, 30(12): 3771–3781.

Zhao XL, Han LH, 2006. Double skin composite construction. *Progress in Structural Engineering and Materials*, 8(3): 93–102.

Appendix 1: Calculation of the Shrinkage of the Sandwiched Concrete

The shrinkage deformation of the concrete in CFDST can be calculated from:

$$\left(\varepsilon_{sh}\right)_t = \frac{t}{35+t} \cdot \left(\varepsilon_{sh}\right)_u \qquad (A1.1)$$

where:

t is the time for concrete to dry (days)

$\left(\varepsilon_{sh}\right)_u$ is the terminal shrinkage of concrete (10^{-6}), calculated from:

$$\left(\varepsilon_{sh}\right)_u = 780 \cdot \gamma_{cp} \cdot \gamma_\lambda \cdot \gamma_{vs} \cdot \gamma_s \cdot \gamma_\omega \cdot \gamma_c \cdot \gamma_\alpha \cdot \gamma_u \cdot \qquad (A1.2)$$

where:

γ_{cp} is the factor of curing time before drying; it can be calculated from Table A1.1 by linear interpolation

γ_λ is the factor of environmental relative humidity, which should be taken as 0.3 for all sandwiched concrete

γ_{vs} is the factor of size influence, which is a function of the ratio of the volume and the superficial area of the component, and can be calculated from:

$$\gamma_{VS} = 1.2 \cdot e^{-0.00472V/S} \tag{A1.3}$$

γ_s is the factor of concrete slump (s, mm), calculated from:

$$\gamma_s = 0.89 + 0.00161s \tag{A1.4}$$

γ_ω is the correction coefficient of fine aggregate, calculated from:

$$\gamma_\omega = 0.30 + 0.014\omega \quad (\omega \le 50) \tag{A1.5}$$

$$\gamma_\omega = 0.90 + 0.002\omega \quad (\omega > 50) \tag{A1.6}$$

where:
  ω is the percentage of fine aggregate
  γ_c is the correction coefficient of the amount of cement, calculated from:

$$\gamma_c = 0.75 + 0.000611c \tag{A1.7}$$

where:
  c is the amount of cement in concrete per cubic meter, kg/m³
  γ_α is the factor of air content in concrete, calculated from:

$$\gamma_\alpha = 0.95 + 0.008\alpha \tag{A1.8}$$

TABLE A1.1 The Influence Coefficient γ_{cp} of the Curing Time before Drying to the Concrete Shrinkage

Curing Time (days)	1	3	7	14	28	90
γ_{cp}	1.20	1.10	1.00	0.93	0.86	0.75

α   is the percentage of air content

γ_u is the factor of steel tube on shrinkage of concrete, calculated from:

$$\gamma_u = 0.0002 D_{size} + 0.63 \tag{A1.9}$$

where:

D_{size} is the size of cross section, $D_{size} = D_o$.

Appendix 2: Stability Coefficient of Members under Axial Compression

TABLE A2.1 Stability Factor χ of Axial Compression Member for Circular Concrete-Filled Double Skin Steel Tube

Steel	Concrete	ψ	α_n	λ											
				10	20	30	40	50	60	70	80	90	100	110	120
S235	C25/30	0	0.05	1.000	0.964	0.910	0.858	0.808	0.761	0.715	0.671	0.629	0.589	0.551	0.495
			0.10	1.000	0.966	0.917	0.869	0.822	0.777	0.733	0.690	0.649	0.609	0.571	0.512
			0.15	1.000	0.968	0.921	0.875	0.831	0.787	0.744	0.702	0.661	0.621	0.582	0.523
			0.20	1.000	0.969	0.924	0.880	0.837	0.794	0.752	0.711	0.670	0.630	0.591	0.530
		0.15	0.05	1.000	0.964	0.910	0.858	0.808	0.761	0.715	0.671	0.629	0.589	0.551	0.492
			0.10	1.000	0.966	0.917	0.869	0.822	0.777	0.733	0.690	0.649	0.609	0.571	0.510
			0.15	1.000	0.968	0.921	0.875	0.831	0.787	0.744	0.702	0.661	0.621	0.582	0.520
			0.20	1.000	0.969	0.924	0.880	0.837	0.794	0.752	0.711	0.670	0.630	0.591	0.528
		0.3	0.05	1.000	0.963	0.910	0.858	0.808	0.760	0.715	0.671	0.629	0.589	0.551	0.485
			0.10	1.000	0.965	0.916	0.868	0.822	0.777	0.733	0.690	0.649	0.609	0.571	0.502
			0.15	1.000	0.967	0.920	0.875	0.830	0.787	0.744	0.702	0.661	0.621	0.582	0.512
			0.20	1.000	0.968	0.924	0.880	0.836	0.794	0.752	0.711	0.670	0.630	0.591	0.519
		0.45	0.05	1.000	0.962	0.909	0.857	0.808	0.760	0.714	0.671	0.629	0.589	0.551	0.472
			0.10	1.000	0.964	0.915	0.868	0.821	0.776	0.733	0.690	0.649	0.609	0.571	0.488
			0.15	1.000	0.966	0.919	0.874	0.830	0.786	0.744	0.702	0.661	0.621	0.582	0.498
			0.20	1.000	0.967	0.923	0.879	0.836	0.793	0.752	0.710	0.670	0.630	0.591	0.506
		0.6	0.05	1.000	0.960	0.907	0.856	0.807	0.759	0.714	0.670	0.629	0.589	0.551	0.454
			0.10	1.000	0.962	0.914	0.866	0.820	0.776	0.732	0.690	0.649	0.609	0.571	0.470

(Continued)

TABLE A2.1 (CONTINUED) Stability Factor χ of Axial Compression Member for Circular Concrete-Filled Double Skin Steel Tube

Steel	Concrete	ψ	α_n	λ											
				10	20	30	40	50	60	70	80	90	100	110	120
			0.15	1.000	0.964	0.918	0.873	0.829	0.786	0.743	0.702	0.661	0.621	0.582	0.480
			0.20	1.000	0.965	0.921	0.878	0.835	0.793	0.751	0.710	0.670	0.630	0.591	0.486
		0.75	0.05	1.000	0.957	0.904	0.854	0.805	0.758	0.713	0.670	0.628	0.589	0.551	0.431
			0.10	1.000	0.958	0.910	0.864	0.819	0.774	0.731	0.689	0.649	0.609	0.571	0.470
			0.15	1.000	0.960	0.915	0.871	0.827	0.784	0.742	0.701	0.661	0.621	0.582	0.480
			0.20	1.000	0.962	0.918	0.876	0.833	0.792	0.750	0.710	0.670	0.630	0.591	0.486
S235	C40/50	0	0.05	1.000	0.937	0.872	0.812	0.754	0.701	0.651	0.605	0.562	0.523	0.488	0.438
			0.10	1.000	0.940	0.879	0.822	0.767	0.716	0.667	0.622	0.580	0.541	0.505	0.454
			0.15	1.000	0.942	0.884	0.828	0.775	0.725	0.677	0.633	0.591	0.552	0.516	0.463
			0.20	1.000	0.944	0.887	0.832	0.780	0.731	0.685	0.640	0.599	0.560	0.523	0.470
		0.15	0.05	1.000	0.937	0.872	0.812	0.754	0.701	0.651	0.605	0.562	0.523	0.488	0.436
			0.10	1.000	0.940	0.879	0.822	0.767	0.716	0.667	0.622	0.580	0.541	0.505	0.451
			0.15	1.000	0.942	0.884	0.828	0.775	0.725	0.677	0.633	0.591	0.552	0.516	0.461
			0.20	1.000	0.943	0.887	0.832	0.780	0.731	0.684	0.640	0.599	0.560	0.523	0.467
		0.3	0.05	1.000	0.937	0.872	0.811	0.754	0.701	0.651	0.605	0.562	0.523	0.488	0.429
			0.10	1.000	0.940	0.879	0.821	0.767	0.716	0.667	0.622	0.580	0.541	0.505	0.444
			0.15	1.000	0.942	0.883	0.827	0.775	0.725	0.677	0.633	0.591	0.552	0.516	0.453
			0.20	1.000	0.943	0.886	0.832	0.780	0.731	0.684	0.640	0.599	0.560	0.523	0.460

(Continued)

TABLE A2.1 (CONTINUED) Stability Factor χ of Axial Compression Member for Circular Concrete-Filled Double Skin Steel Tube

Steel	Concrete	ψ	α_n	λ											
				10	20	30	40	50	60	70	80	90	100	110	120
		0.45	0.05	1.000	0.936	0.872	0.811	0.754	0.701	0.651	0.605	0.562	0.523	0.488	0.418
			0.10	1.000	0.939	0.878	0.821	0.767	0.715	0.667	0.622	0.580	0.541	0.505	0.432
			0.15	1.000	0.941	0.882	0.827	0.774	0.724	0.677	0.633	0.591	0.552	0.516	0.441
			0.20	1.000	0.942	0.885	0.831	0.780	0.731	0.684	0.640	0.599	0.560	0.523	0.448
		0.6	0.05	1.000	0.935	0.871	0.810	0.753	0.700	0.650	0.604	0.562	0.523	0.488	0.402
			0.10	1.000	0.937	0.877	0.820	0.766	0.715	0.667	0.622	0.580	0.541	0.505	0.416
			0.15	1.000	0.939	0.881	0.826	0.773	0.724	0.677	0.632	0.591	0.552	0.516	0.425
			0.20	0.999	0.940	0.884	0.830	0.779	0.730	0.684	0.640	0.599	0.560	0.523	0.431
		0.75	0.05	0.999	0.932	0.869	0.809	0.752	0.699	0.650	0.604	0.562	0.523	0.488	0.381
			0.10	0.997	0.934	0.875	0.818	0.764	0.714	0.666	0.622	0.580	0.541	0.505	0.416
			0.15	0.996	0.936	0.879	0.824	0.772	0.723	0.676	0.632	0.591	0.552	0.516	0.425
			0.20	0.996	0.938	0.882	0.829	0.778	0.729	0.683	0.640	0.598	0.560	0.523	0.431
S235	C55/67	0	0.05	0.994	0.918	0.847	0.781	0.719	0.662	0.610	0.562	0.520	0.481	0.448	0.402
			0.10	0.993	0.922	0.854	0.790	0.731	0.676	0.625	0.578	0.536	0.498	0.464	0.416
			0.15	0.993	0.924	0.858	0.796	0.738	0.684	0.634	0.588	0.546	0.508	0.473	0.425
			0.20	0.993	0.925	0.861	0.800	0.743	0.690	0.641	0.595	0.553	0.515	0.480	0.431
		0.15	0.05	0.994	0.918	0.847	0.781	0.719	0.662	0.610	0.562	0.520	0.481	0.448	0.400
			0.10	0.993	0.921	0.854	0.790	0.731	0.676	0.625	0.578	0.536	0.498	0.464	0.414

(Continued)

TABLE A2.1 (CONTINUED) Stability Factor χ of Axial Compression Member for Circular Concrete-Filled Double Skin Steel Tube

Steel	Concrete	ψ	α_n	λ												
				10	20	30	40	50	60	70	80	90	100	110	120
		0.3	0.15	0.993	0.924	0.858	0.796	0.738	0.684	0.634	0.588	0.546	0.508	0.473	0.422
			0.20	0.993	0.925	0.861	0.800	0.743	0.690	0.641	0.595	0.553	0.515	0.480	0.429
			0.05	0.993	0.918	0.847	0.780	0.719	0.662	0.610	0.562	0.519	0.481	0.448	0.394
			0.10	0.993	0.921	0.854	0.790	0.731	0.676	0.625	0.578	0.536	0.498	0.464	0.408
			0.15	0.993	0.923	0.858	0.796	0.738	0.684	0.634	0.588	0.546	0.508	0.473	0.416
			0.20	0.992	0.925	0.861	0.800	0.743	0.690	0.641	0.595	0.553	0.515	0.480	0.422
		0.45	0.05	0.993	0.917	0.846	0.780	0.719	0.662	0.610	0.562	0.519	0.481	0.448	0.383
			0.10	0.992	0.921	0.853	0.790	0.731	0.676	0.625	0.578	0.536	0.498	0.464	0.397
			0.15	0.992	0.923	0.857	0.795	0.738	0.684	0.634	0.588	0.546	0.508	0.473	0.405
			0.20	0.992	0.924	0.860	0.800	0.743	0.690	0.641	0.595	0.553	0.515	0.480	0.411
		0.6	0.05	0.992	0.916	0.846	0.780	0.718	0.662	0.610	0.562	0.519	0.481	0.448	0.369
			0.10	0.991	0.919	0.852	0.789	0.730	0.675	0.625	0.578	0.536	0.498	0.464	0.382
			0.15	0.990	0.921	0.856	0.795	0.737	0.684	0.634	0.588	0.546	0.508	0.473	0.389
			0.20	0.990	0.923	0.859	0.799	0.742	0.690	0.640	0.595	0.553	0.515	0.480	0.395
		0.75	0.05	0.990	0.915	0.844	0.779	0.718	0.661	0.609	0.562	0.519	0.481	0.448	0.350
			0.10	0.988	0.917	0.850	0.788	0.729	0.675	0.624	0.578	0.536	0.498	0.464	0.382
			0.15	0.988	0.919	0.854	0.793	0.736	0.683	0.633	0.588	0.546	0.507	0.473	0.389
			0.20	0.988	0.921	0.857	0.798	0.741	0.689	0.640	0.595	0.553	0.515	0.480	0.395

(Continued)

TABLE A2.1 (CONTINUED) Stability Factor χ of Axial Compression Member for Circular Concrete-Filled Double Skin Steel Tube

Steel	Concrete	ψ	α_n	λ											
				10	20	30	40	50	60	70	80	90	100	110	120
S235	C70/85	0	0.05	0.985	0.904	0.828	0.758	0.693	0.634	0.580	0.532	0.489	0.451	0.419	0.376
			0.10	0.985	0.907	0.835	0.767	0.705	0.647	0.595	0.547	0.504	0.466	0.434	0.389
			0.15	0.985	0.909	0.839	0.773	0.712	0.655	0.603	0.556	0.513	0.476	0.443	0.397
			0.20	0.985	0.911	0.842	0.777	0.717	0.661	0.609	0.563	0.520	0.482	0.449	0.403
		0.15	0.05	0.985	0.904	0.828	0.758	0.693	0.634	0.580	0.532	0.489	0.451	0.419	0.374
			0.10	0.985	0.907	0.835	0.767	0.705	0.647	0.595	0.547	0.504	0.466	0.434	0.387
			0.15	0.985	0.909	0.839	0.773	0.712	0.655	0.603	0.556	0.513	0.476	0.443	0.395
			0.20	0.985	0.911	0.842	0.777	0.716	0.661	0.609	0.562	0.520	0.482	0.449	0.401
		0.3	0.05	0.984	0.903	0.828	0.758	0.693	0.634	0.580	0.532	0.489	0.451	0.419	0.368
			0.10	0.984	0.907	0.835	0.767	0.705	0.647	0.594	0.547	0.504	0.466	0.434	0.381
			0.15	0.984	0.909	0.839	0.773	0.711	0.655	0.603	0.556	0.513	0.476	0.443	0.389
			0.20	0.984	0.911	0.841	0.777	0.716	0.661	0.609	0.562	0.520	0.482	0.449	0.395
		0.45	0.05	0.984	0.903	0.828	0.758	0.693	0.634	0.580	0.532	0.489	0.451	0.419	0.358
			0.10	0.984	0.907	0.834	0.767	0.704	0.647	0.594	0.547	0.504	0.466	0.434	0.371
			0.15	0.984	0.909	0.838	0.772	0.711	0.655	0.603	0.556	0.513	0.476	0.443	0.379
			0.20	0.984	0.910	0.841	0.776	0.716	0.660	0.609	0.562	0.520	0.482	0.449	0.384
		0.6	0.05	0.983	0.903	0.827	0.757	0.693	0.634	0.580	0.532	0.489	0.451	0.419	0.345
			0.10	0.983	0.906	0.834	0.766	0.704	0.647	0.594	0.547	0.504	0.466	0.434	0.357

(Continued)

TABLE A2.1 (CONTINUED) Stability Factor χ of Axial Compression Member for Circular Concrete-Filled Double Skin Steel Tube

Steel	Concrete	ψ	α_n	λ											
				10	20	30	40	50	60	70	80	90	100	110	120
		0.75	0.15	0.983	0.908	0.837	0.772	0.711	0.654	0.603	0.556	0.513	0.476	0.443	0.364
			0.20	0.983	0.909	0.840	0.776	0.716	0.660	0.609	0.562	0.520	0.482	0.449	0.370
			0.05	0.982	0.901	0.826	0.757	0.692	0.633	0.580	0.531	0.489	0.451	0.419	0.327
			0.10	0.981	0.904	0.832	0.765	0.703	0.646	0.594	0.546	0.504	0.466	0.434	0.357
			0.15	0.981	0.906	0.836	0.771	0.710	0.654	0.602	0.556	0.513	0.476	0.443	0.364
			0.20	0.981	0.908	0.839	0.775	0.715	0.660	0.609	0.562	0.520	0.482	0.449	0.370
S275	C25/30	0	0.05	1.000	0.966	0.915	0.865	0.816	0.769	0.723	0.678	0.635	0.593	0.534	0.467
			0.10	1.000	0.969	0.923	0.877	0.832	0.787	0.743	0.699	0.656	0.613	0.552	0.483
			0.15	1.000	0.971	0.928	0.884	0.841	0.798	0.755	0.712	0.669	0.626	0.564	0.493
			0.20	1.000	0.972	0.931	0.890	0.848	0.806	0.764	0.721	0.678	0.635	0.572	0.501
		0.15	0.05	1.000	0.966	0.914	0.865	0.816	0.769	0.723	0.678	0.635	0.593	0.531	0.465
			0.10	1.000	0.969	0.922	0.877	0.832	0.787	0.743	0.699	0.656	0.613	0.550	0.481
			0.15	1.000	0.970	0.927	0.884	0.841	0.798	0.755	0.712	0.669	0.626	0.561	0.491
			0.20	1.000	0.972	0.931	0.890	0.848	0.806	0.764	0.721	0.678	0.635	0.569	0.498
		0.3	0.05	1.000	0.965	0.914	0.864	0.816	0.769	0.723	0.678	0.635	0.593	0.523	0.457
			0.10	1.000	0.968	0.922	0.876	0.831	0.787	0.743	0.699	0.656	0.613	0.541	0.473
			0.15	1.000	0.970	0.927	0.884	0.841	0.798	0.755	0.712	0.669	0.626	0.552	0.483
			0.20	1.000	0.971	0.930	0.889	0.848	0.806	0.764	0.721	0.678	0.635	0.560	0.490

(Continued)

TABLE A2.1 (CONTINUED) Stability Factor χ of Axial Compression Member for Circular Concrete-Filled Double Skin Steel Tube

Steel	Concrete	ψ	α_n	λ											
				10	20	30	40	50	60	70	80	90	100	110	120
		0.45	0.05	1.000	0.964	0.913	0.864	0.815	0.768	0.722	0.678	0.635	0.593	0.509	0.445
			0.10	1.000	0.967	0.921	0.876	0.831	0.787	0.743	0.699	0.656	0.613	0.527	0.461
			0.15	1.000	0.969	0.926	0.883	0.840	0.798	0.755	0.712	0.669	0.626	0.537	0.470
			0.20	1.000	0.970	0.930	0.889	0.847	0.806	0.764	0.721	0.678	0.635	0.545	0.477
		0.6	0.05	1.000	0.962	0.912	0.862	0.814	0.768	0.722	0.678	0.635	0.593	0.489	0.428
			0.10	1.000	0.964	0.919	0.874	0.830	0.786	0.742	0.699	0.656	0.613	0.507	0.443
			0.15	1.000	0.967	0.924	0.882	0.840	0.797	0.754	0.712	0.669	0.626	0.517	0.453
			0.20	1.000	0.968	0.928	0.888	0.847	0.805	0.763	0.721	0.678	0.635	0.525	0.459
		0.75	0.05	1.000	0.958	0.909	0.860	0.813	0.767	0.721	0.677	0.634	0.593	0.465	0.407
			0.10	1.000	0.961	0.916	0.872	0.828	0.785	0.742	0.699	0.656	0.613	0.507	0.443
			0.15	1.000	0.963	0.922	0.880	0.838	0.796	0.754	0.711	0.669	0.626	0.517	0.453
			0.20	1.000	0.965	0.926	0.886	0.845	0.804	0.763	0.721	0.678	0.635	0.525	0.459
S275	C40/50	0	0.05	1.000	0.938	0.874	0.814	0.757	0.704	0.654	0.608	0.565	0.525	0.473	0.414
			0.10	1.000	0.942	0.882	0.826	0.772	0.721	0.672	0.626	0.584	0.543	0.489	0.428
			0.15	1.000	0.944	0.887	0.833	0.780	0.731	0.683	0.638	0.595	0.554	0.499	0.437
			0.20	1.000	0.946	0.891	0.838	0.787	0.738	0.691	0.646	0.603	0.562	0.506	0.443
		0.15	0.05	1.000	0.938	0.874	0.814	0.757	0.704	0.654	0.608	0.565	0.525	0.470	0.411
			0.10	1.000	0.942	0.882	0.826	0.772	0.721	0.672	0.626	0.584	0.543	0.487	0.426

(Continued)

TABLE A2.1 (CONTINUED)   Stability Factor χ of Axial Compression Member for Circular Concrete-Filled Double Skin Steel Tube

Steel	Concrete	ψ	α_n	10	20	30	40	50	60	70	80	90	100	110	120
			0.15	1.000	0.944	0.887	0.832	0.780	0.730	0.683	0.638	0.595	0.554	0.497	0.435
			0.20	1.000	0.945	0.890	0.838	0.787	0.738	0.691	0.646	0.603	0.562	0.504	0.441
		0.3	0.05	1.000	0.937	0.874	0.814	0.757	0.704	0.654	0.608	0.565	0.525	0.463	0.405
			0.10	1.000	0.941	0.882	0.825	0.771	0.720	0.672	0.626	0.584	0.543	0.479	0.419
			0.15	1.000	0.943	0.887	0.832	0.780	0.730	0.683	0.638	0.595	0.554	0.489	0.428
			0.20	1.000	0.945	0.890	0.837	0.786	0.738	0.691	0.646	0.603	0.562	0.496	0.434
		0.45	0.05	1.000	0.937	0.873	0.813	0.757	0.704	0.654	0.607	0.565	0.525	0.450	0.394
			0.10	1.000	0.940	0.881	0.825	0.771	0.720	0.672	0.626	0.584	0.543	0.466	0.408
			0.15	1.000	0.942	0.886	0.832	0.780	0.730	0.683	0.638	0.595	0.554	0.476	0.417
			0.20	1.000	0.944	0.889	0.837	0.786	0.737	0.691	0.646	0.603	0.562	0.483	0.423
		0.6	0.05	1.000	0.935	0.872	0.813	0.756	0.703	0.654	0.607	0.564	0.525	0.433	0.379
			0.10	1.000	0.938	0.880	0.824	0.770	0.720	0.672	0.626	0.583	0.543	0.449	0.393
			0.15	0.999	0.941	0.884	0.831	0.779	0.730	0.682	0.638	0.595	0.554	0.458	0.401
			0.20	0.999	0.942	0.888	0.836	0.785	0.737	0.690	0.646	0.603	0.562	0.464	0.406
		0.75	0.05	0.998	0.933	0.870	0.811	0.755	0.702	0.653	0.607	0.564	0.525	0.411	0.360
			0.10	0.996	0.935	0.877	0.822	0.769	0.719	0.671	0.626	0.583	0.543	0.449	0.393
			0.15	0.995	0.938	0.882	0.829	0.778	0.729	0.682	0.637	0.595	0.554	0.458	0.401
			0.20	0.995	0.940	0.886	0.834	0.784	0.736	0.690	0.646	0.603	0.562	0.464	0.406

(Continued)

TABLE A2.1 (CONTINUED) Stability Factor χ of Axial Compression Member for Circular Concrete-Filled Double Skin Steel Tube

Steel	Concrete	ψ	α_n	λ											
				10	20	30	40	50	60	70	80	90	100	110	120
S275	C55/67	0	0.05	0.994	0.918	0.847	0.780	0.719	0.662	0.610	0.562	0.520	0.482	0.433	0.379
			0.10	0.993	0.922	0.855	0.791	0.732	0.677	0.626	0.580	0.537	0.499	0.449	0.393
			0.15	0.993	0.924	0.859	0.798	0.740	0.687	0.636	0.590	0.548	0.509	0.458	0.401
			0.20	0.993	0.926	0.863	0.803	0.746	0.693	0.644	0.598	0.555	0.516	0.465	0.407
		0.15	0.05	0.993	0.918	0.847	0.780	0.719	0.662	0.610	0.562	0.520	0.482	0.431	0.377
			0.10	0.993	0.922	0.854	0.791	0.732	0.677	0.626	0.580	0.537	0.499	0.446	0.391
			0.15	0.993	0.924	0.859	0.798	0.740	0.686	0.636	0.590	0.548	0.509	0.456	0.399
			0.20	0.993	0.926	0.863	0.803	0.746	0.693	0.644	0.598	0.555	0.516	0.462	0.404
		0.3	0.05	0.993	0.917	0.846	0.780	0.719	0.662	0.610	0.562	0.520	0.482	0.424	0.371
			0.10	0.993	0.921	0.854	0.791	0.732	0.677	0.626	0.580	0.537	0.499	0.439	0.385
			0.15	0.992	0.924	0.859	0.798	0.740	0.686	0.636	0.590	0.548	0.509	0.448	0.392
			0.20	0.992	0.926	0.862	0.802	0.746	0.693	0.644	0.598	0.555	0.516	0.455	0.398
		0.45	0.05	0.992	0.917	0.846	0.780	0.718	0.662	0.610	0.562	0.520	0.482	0.413	0.362
			0.10	0.992	0.921	0.854	0.791	0.732	0.677	0.626	0.580	0.537	0.499	0.428	0.374
			0.15	0.992	0.923	0.858	0.797	0.740	0.686	0.636	0.590	0.548	0.509	0.437	0.382
			0.20	0.991	0.925	0.862	0.802	0.746	0.693	0.644	0.598	0.555	0.516	0.443	0.388
		0.6	0.05	0.991	0.916	0.845	0.779	0.718	0.661	0.609	0.562	0.520	0.482	0.398	0.348
			0.10	0.990	0.919	0.853	0.790	0.731	0.677	0.626	0.580	0.537	0.499	0.412	0.360

(Continued)

TABLE A2.1 (CONTINUED) Stability Factor χ of Axial Compression Member for Circular Concrete-Filled Double Skin Steel Tube

Steel	Concrete	ψ	α_n	λ											
				10	20	30	40	50	60	70	80	90	100	110	120
		0.75	0.15	0.990	0.922	0.857	0.796	0.739	0.686	0.636	0.590	0.548	0.509	0.420	0.368
			0.20	0.990	0.924	0.861	0.801	0.745	0.693	0.643	0.597	0.555	0.516	0.426	0.373
S275	C70/85	0	0.05	0.989	0.914	0.844	0.778	0.717	0.661	0.609	0.562	0.520	0.482	0.377	0.330
			0.10	0.987	0.917	0.851	0.789	0.730	0.676	0.626	0.579	0.537	0.499	0.412	0.360
			0.15	0.987	0.919	0.855	0.795	0.738	0.685	0.636	0.590	0.547	0.509	0.420	0.368
			0.20	0.987	0.921	0.859	0.800	0.744	0.692	0.643	0.597	0.555	0.516	0.426	0.373
		0.15	0.05	0.984	0.902	0.826	0.756	0.691	0.631	0.578	0.530	0.487	0.451	0.405	0.355
			0.10	0.984	0.907	0.834	0.766	0.703	0.646	0.593	0.546	0.504	0.467	0.420	0.367
			0.15	0.985	0.909	0.838	0.772	0.711	0.655	0.603	0.556	0.514	0.476	0.428	0.375
			0.20	0.985	0.911	0.842	0.777	0.717	0.661	0.610	0.563	0.521	0.483	0.434	0.380
		0.3	0.05	0.984	0.902	0.826	0.755	0.691	0.631	0.578	0.530	0.487	0.451	0.397	0.347
			0.10	0.984	0.906	0.834	0.766	0.703	0.646	0.593	0.546	0.504	0.467	0.411	0.360
			0.15	0.984	0.909	0.838	0.772	0.711	0.655	0.603	0.556	0.514	0.476	0.419	0.367
			0.20	0.984	0.911	0.841	0.777	0.717	0.661	0.610	0.563	0.521	0.483	0.425	0.372

(Continued)

TABLE A2.1 (CONTINUED) Stability Factor χ of Axial Compression Member for Circular Concrete-Filled Double Skin Steel Tube

Steel	Concrete	ψ	α_n	λ											
				10	20	30	40	50	60	70	80	90	100	110	120
		0.45	0.05	0.984	0.902	0.826	0.755	0.690	0.631	0.578	0.530	0.487	0.451	0.386	0.338
			0.10	0.983	0.906	0.833	0.766	0.703	0.646	0.593	0.546	0.504	0.467	0.400	0.350
			0.15	0.983	0.908	0.838	0.772	0.711	0.654	0.603	0.556	0.514	0.476	0.408	0.357
			0.20	0.983	0.910	0.841	0.776	0.716	0.661	0.610	0.563	0.521	0.483	0.414	0.363
		0.6	0.05	0.983	0.901	0.825	0.755	0.690	0.631	0.578	0.530	0.487	0.451	0.372	0.325
			0.10	0.982	0.905	0.832	0.765	0.703	0.645	0.593	0.546	0.504	0.467	0.385	0.337
			0.15	0.982	0.907	0.837	0.771	0.710	0.654	0.603	0.556	0.514	0.476	0.393	0.344
			0.20	0.982	0.909	0.840	0.776	0.716	0.660	0.609	0.563	0.521	0.483	0.398	0.349
		0.75	0.05	0.981	0.900	0.824	0.754	0.690	0.631	0.577	0.530	0.487	0.451	0.353	0.309
			0.10	0.980	0.903	0.831	0.764	0.702	0.645	0.593	0.546	0.504	0.467	0.385	0.337
			0.15	0.980	0.905	0.835	0.770	0.710	0.654	0.602	0.556	0.513	0.476	0.393	0.344
			0.20	0.980	0.907	0.839	0.775	0.715	0.660	0.609	0.563	0.521	0.483	0.398	0.349
S355	C25/30	0	0.05	1.000	0.969	0.923	0.875	0.828	0.780	0.732	0.684	0.635	0.556	0.482	0.422
			0.10	1.000	0.973	0.933	0.891	0.847	0.802	0.755	0.707	0.658	0.576	0.499	0.437
			0.15	1.000	0.976	0.939	0.900	0.858	0.815	0.769	0.721	0.671	0.587	0.509	0.446
			0.20	1.000	0.978	0.943	0.906	0.866	0.824	0.779	0.731	0.681	0.596	0.516	0.452
		0.15	0.05	1.000	0.969	0.922	0.875	0.828	0.780	0.732	0.684	0.635	0.553	0.479	0.420
			0.10	1.000	0.973	0.932	0.890	0.847	0.802	0.755	0.707	0.658	0.573	0.496	0.434

(Continued)

TABLE A2.1 (CONTINUED) Stability Factor χ of Axial Compression Member for Circular Concrete-Filled Double Skin Steel Tube

Steel	Concrete	ψ	α_n	λ											
				10	20	30	40	50	60	70	80	90	100	110	120
			0.15	1.000	0.976	0.939	0.900	0.858	0.815	0.769	0.721	0.671	0.584	0.506	0.443
			0.20	1.000	0.978	0.943	0.906	0.866	0.824	0.779	0.731	0.681	0.593	0.514	0.450
		0.3	0.05	1.000	0.969	0.922	0.875	0.828	0.780	0.732	0.684	0.635	0.544	0.472	0.413
			0.10	1.000	0.972	0.932	0.890	0.847	0.802	0.755	0.707	0.658	0.564	0.489	0.428
			0.15	1.000	0.975	0.938	0.899	0.858	0.815	0.769	0.721	0.671	0.575	0.499	0.436
			0.20	1.000	0.977	0.943	0.906	0.866	0.824	0.779	0.731	0.681	0.583	0.506	0.443
		0.45	0.05	1.000	0.967	0.921	0.874	0.827	0.780	0.732	0.684	0.635	0.530	0.459	0.402
			0.10	1.000	0.971	0.931	0.889	0.846	0.801	0.755	0.707	0.658	0.549	0.476	0.416
			0.15	1.000	0.974	0.937	0.899	0.858	0.814	0.769	0.721	0.671	0.560	0.485	0.425
			0.20	1.000	0.976	0.942	0.905	0.866	0.824	0.779	0.731	0.681	0.568	0.492	0.431
		0.6	0.05	1.000	0.965	0.920	0.873	0.827	0.780	0.732	0.684	0.635	0.510	0.442	0.387
			0.10	1.000	0.969	0.930	0.888	0.846	0.801	0.755	0.707	0.658	0.528	0.458	0.400
			0.15	1.000	0.972	0.936	0.898	0.857	0.814	0.769	0.721	0.671	0.539	0.467	0.409
			0.20	1.000	0.975	0.941	0.905	0.865	0.823	0.779	0.731	0.681	0.546	0.474	0.415
		0.75	0.05	1.000	0.962	0.917	0.872	0.826	0.779	0.732	0.684	0.635	0.484	0.420	0.367
			0.10	1.000	0.966	0.927	0.887	0.844	0.800	0.755	0.707	0.658	0.528	0.458	0.400
			0.15	1.000	0.970	0.934	0.896	0.856	0.814	0.768	0.721	0.671	0.539	0.467	0.409
			0.20	1.000	0.972	0.939	0.903	0.865	0.823	0.779	0.731	0.681	0.546	0.474	0.415

(Continued)

TABLE A2.1 (CONTINUED) Stability Factor χ of Axial Compression Member for Circular Concrete-Filled Double Skin Steel Tube

Steel	Concrete	ψ	α_n	λ											
				10	20	30	40	50	60	70	80	90	100	110	120
S355	C40/50	0	0.05	1.000	0.939	0.877	0.818	0.761	0.707	0.656	0.608	0.563	0.492	0.427	0.373
			0.10	1.000	0.944	0.887	0.832	0.778	0.727	0.677	0.629	0.583	0.510	0.442	0.387
			0.15	1.000	0.947	0.893	0.840	0.789	0.738	0.689	0.641	0.594	0.520	0.451	0.395
			0.20	1.000	0.949	0.897	0.846	0.796	0.747	0.698	0.650	0.603	0.528	0.457	0.400
		0.15	0.05	1.000	0.939	0.877	0.818	0.761	0.707	0.656	0.608	0.563	0.490	0.424	0.371
			0.10	1.000	0.944	0.887	0.832	0.778	0.727	0.677	0.629	0.583	0.507	0.439	0.385
			0.15	1.000	0.947	0.893	0.840	0.789	0.738	0.689	0.641	0.594	0.517	0.448	0.392
			0.20	1.000	0.949	0.897	0.846	0.796	0.747	0.698	0.650	0.603	0.525	0.455	0.398
		0.3	0.05	1.000	0.939	0.877	0.817	0.761	0.707	0.656	0.608	0.563	0.482	0.418	0.366
			0.10	1.000	0.944	0.887	0.831	0.778	0.726	0.677	0.629	0.583	0.499	0.433	0.379
			0.15	1.000	0.947	0.893	0.840	0.788	0.738	0.689	0.641	0.594	0.509	0.441	0.386
			0.20	1.000	0.949	0.897	0.846	0.796	0.746	0.698	0.650	0.603	0.517	0.448	0.392
		0.45	0.05	1.000	0.938	0.876	0.817	0.761	0.707	0.656	0.608	0.563	0.469	0.407	0.356
			0.10	1.000	0.943	0.886	0.831	0.778	0.726	0.677	0.629	0.583	0.486	0.421	0.369
			0.15	1.000	0.945	0.892	0.839	0.788	0.738	0.689	0.641	0.594	0.496	0.430	0.376
			0.20	1.000	0.948	0.896	0.845	0.795	0.746	0.698	0.650	0.603	0.503	0.436	0.382
		0.6	0.05	1.000	0.936	0.875	0.816	0.760	0.707	0.656	0.608	0.563	0.451	0.391	0.343
			0.10	0.999	0.941	0.884	0.830	0.777	0.726	0.676	0.629	0.583	0.467	0.405	0.355

(Continued)

TABLE A2.1 (CONTINUED) Stability Factor χ of Axial Compression Member for Circular Concrete-Filled Double Skin Steel Tube

Steel	Concrete	ψ	α_n	λ											
				10	20	30	40	50	60	70	80	90	100	110	120
			0.15	0.998	0.944	0.891	0.838	0.787	0.737	0.689	0.641	0.594	0.477	0.413	0.362
			0.20	0.998	0.946	0.895	0.845	0.795	0.746	0.698	0.650	0.603	0.484	0.419	0.367
		0.75	0.05	0.997	0.933	0.873	0.815	0.759	0.706	0.656	0.608	0.563	0.429	0.371	0.325
			0.10	0.995	0.938	0.882	0.828	0.776	0.725	0.676	0.629	0.583	0.467	0.405	0.355
			0.15	0.994	0.941	0.888	0.837	0.786	0.737	0.688	0.641	0.594	0.477	0.413	0.362
			0.20	0.995	0.944	0.893	0.843	0.794	0.745	0.697	0.650	0.603	0.484	0.419	0.367
S355	C55/67	0	0.05	0.993	0.917	0.846	0.779	0.717	0.660	0.607	0.559	0.516	0.452	0.391	0.343
			0.10	0.993	0.922	0.856	0.792	0.733	0.678	0.626	0.578	0.534	0.467	0.405	0.355
			0.15	0.993	0.925	0.861	0.800	0.743	0.689	0.638	0.590	0.545	0.477	0.414	0.362
			0.20	0.993	0.928	0.865	0.806	0.750	0.696	0.646	0.598	0.553	0.484	0.420	0.367
		0.15	0.05	0.993	0.917	0.846	0.779	0.717	0.660	0.607	0.559	0.516	0.449	0.389	0.341
			0.10	0.993	0.922	0.855	0.792	0.733	0.678	0.626	0.578	0.534	0.465	0.403	0.353
			0.15	0.993	0.925	0.861	0.800	0.743	0.688	0.637	0.590	0.545	0.475	0.411	0.360
			0.20	0.993	0.928	0.865	0.806	0.750	0.696	0.646	0.598	0.553	0.481	0.417	0.365
		0.3	0.05	0.993	0.917	0.846	0.779	0.717	0.660	0.607	0.559	0.516	0.442	0.383	0.335
			0.10	0.992	0.922	0.855	0.792	0.733	0.678	0.626	0.578	0.534	0.458	0.397	0.347
			0.15	0.992	0.925	0.861	0.800	0.743	0.688	0.637	0.590	0.545	0.467	0.405	0.354
			0.20	0.992	0.927	0.865	0.806	0.750	0.696	0.646	0.598	0.553	0.474	0.411	0.360

(Continued)

TABLE A2.1 (CONTINUED) Stability Factor χ of Axial Compression Member for Circular Concrete-Filled Double Skin Steel Tube

Steel	Concrete	ψ	α_n	λ											
				10	20	30	40	50	60	70	80	90	100	110	120
		0.45	0.05	0.992	0.916	0.845	0.779	0.717	0.660	0.607	0.559	0.516	0.431	0.373	0.327
			0.10	0.991	0.921	0.854	0.792	0.733	0.678	0.626	0.578	0.534	0.446	0.386	0.338
			0.15	0.991	0.924	0.860	0.800	0.742	0.688	0.637	0.590	0.545	0.455	0.394	0.345
			0.20	0.991	0.926	0.864	0.805	0.749	0.696	0.646	0.598	0.553	0.461	0.400	0.350
		0.6	0.05	0.990	0.915	0.844	0.778	0.717	0.660	0.607	0.559	0.516	0.414	0.359	0.314
			0.10	0.989	0.920	0.853	0.791	0.732	0.677	0.626	0.578	0.534	0.429	0.372	0.325
			0.15	0.989	0.923	0.859	0.799	0.742	0.688	0.637	0.590	0.545	0.438	0.379	0.332
			0.20	0.989	0.925	0.863	0.805	0.749	0.696	0.645	0.598	0.553	0.444	0.385	0.337
		0.75	0.05	0.987	0.913	0.843	0.777	0.716	0.659	0.607	0.559	0.516	0.393	0.341	0.298
			0.10	0.986	0.917	0.851	0.790	0.731	0.677	0.626	0.578	0.534	0.429	0.372	0.325
			0.15	0.986	0.920	0.857	0.797	0.741	0.687	0.637	0.590	0.545	0.438	0.379	0.332
			0.20	0.986	0.922	0.862	0.803	0.748	0.695	0.645	0.598	0.553	0.444	0.385	0.337
S355	C70/85	0	0.05	0.984	0.900	0.823	0.751	0.685	0.626	0.572	0.525	0.483	0.422	0.366	0.320
			0.10	0.984	0.905	0.832	0.764	0.700	0.643	0.590	0.542	0.500	0.437	0.379	0.332
			0.15	0.984	0.908	0.837	0.771	0.709	0.653	0.600	0.553	0.510	0.446	0.387	0.338
			0.20	0.984	0.911	0.841	0.777	0.716	0.660	0.608	0.561	0.517	0.453	0.392	0.343
		0.15	0.05	0.984	0.900	0.823	0.751	0.685	0.626	0.572	0.525	0.483	0.420	0.364	0.319
			0.10	0.984	0.905	0.832	0.763	0.700	0.642	0.590	0.542	0.500	0.435	0.377	0.330

(Continued)

TABLE A2.1 (CONTINUED) Stability Factor χ of Axial Compression Member for Circular Concrete-Filled Double Skin Steel Tube

Steel	Concrete	ψ	α_n	λ											
				10	20	30	40	50	60	70	80	90	100	110	120
			0.15	0.984	0.908	0.837	0.771	0.709	0.653	0.600	0.553	0.510	0.444	0.385	0.337
			0.20	0.984	0.911	0.841	0.776	0.716	0.660	0.608	0.561	0.517	0.450	0.390	0.342
		0.3	0.05	0.983	0.900	0.822	0.751	0.685	0.626	0.572	0.525	0.483	0.414	0.358	0.314
			0.10	0.984	0.905	0.832	0.763	0.700	0.642	0.590	0.542	0.500	0.428	0.371	0.325
			0.15	0.984	0.908	0.837	0.771	0.709	0.652	0.600	0.553	0.510	0.437	0.379	0.331
			0.20	0.984	0.910	0.841	0.776	0.716	0.660	0.608	0.561	0.517	0.443	0.384	0.336
		0.45	0.05	0.983	0.899	0.822	0.751	0.685	0.626	0.572	0.525	0.483	0.403	0.349	0.305
			0.10	0.983	0.904	0.831	0.763	0.700	0.642	0.590	0.542	0.500	0.417	0.361	0.316
			0.15	0.983	0.907	0.837	0.770	0.709	0.652	0.600	0.553	0.510	0.425	0.369	0.323
			0.20	0.983	0.910	0.841	0.776	0.716	0.660	0.608	0.561	0.517	0.432	0.374	0.327
		0.6	0.05	0.982	0.898	0.821	0.750	0.685	0.625	0.572	0.524	0.483	0.387	0.336	0.294
			0.10	0.981	0.903	0.830	0.762	0.700	0.642	0.590	0.542	0.500	0.401	0.348	0.304
			0.15	0.981	0.906	0.836	0.770	0.709	0.652	0.600	0.553	0.510	0.409	0.355	0.310
			0.20	0.981	0.908	0.840	0.775	0.715	0.659	0.608	0.560	0.517	0.415	0.360	0.315
		0.75	0.05	0.979	0.897	0.820	0.749	0.684	0.625	0.572	0.524	0.483	0.368	0.319	0.279
			0.10	0.978	0.901	0.829	0.761	0.699	0.642	0.589	0.542	0.500	0.401	0.348	0.304
			0.15	0.978	0.904	0.834	0.769	0.708	0.652	0.600	0.553	0.510	0.409	0.355	0.310
			0.20	0.979	0.906	0.838	0.774	0.714	0.659	0.608	0.560	0.517	0.415	0.360	0.315

(Continued)

TABLE A2.1 (CONTINUED)   Stability Factor χ of Axial Compression Member for Circular Concrete-Filled Double Skin Steel Tube

Steel	Concrete	ψ	α_n	λ											
				10	20	30	40	50	60	70	80	90	100	110	120
S450	C25/30	0	0.05	1.000	0.972	0.929	0.883	0.835	0.785	0.732	0.677	0.584	0.501	0.434	0.380
			0.10	1.000	0.977	0.941	0.901	0.857	0.808	0.756	0.701	0.604	0.518	0.449	0.393
			0.15	1.000	0.981	0.948	0.911	0.870	0.823	0.771	0.715	0.617	0.529	0.458	0.401
			0.20	1.000	0.983	0.954	0.919	0.879	0.833	0.782	0.725	0.626	0.536	0.465	0.407
		0.15	0.05	1.000	0.972	0.929	0.883	0.835	0.784	0.732	0.677	0.581	0.498	0.432	0.378
			0.10	1.000	0.977	0.941	0.901	0.857	0.808	0.756	0.701	0.601	0.515	0.447	0.391
			0.15	1.000	0.980	0.948	0.911	0.870	0.823	0.771	0.715	0.614	0.526	0.456	0.399
			0.20	1.000	0.983	0.954	0.919	0.879	0.833	0.782	0.725	0.622	0.534	0.463	0.405
		0.3	0.05	1.000	0.972	0.928	0.883	0.835	0.784	0.732	0.677	0.572	0.490	0.425	0.372
			0.10	1.000	0.977	0.941	0.900	0.856	0.808	0.756	0.701	0.592	0.507	0.440	0.385
			0.15	1.000	0.980	0.948	0.911	0.869	0.823	0.771	0.715	0.604	0.518	0.449	0.393
			0.20	1.000	0.982	0.953	0.919	0.879	0.833	0.782	0.725	0.613	0.525	0.455	0.399
		0.45	0.05	1.000	0.970	0.928	0.882	0.834	0.784	0.732	0.677	0.557	0.477	0.414	0.362
			0.10	1.000	0.976	0.940	0.900	0.856	0.808	0.756	0.701	0.576	0.494	0.428	0.375
			0.15	1.000	0.979	0.947	0.911	0.869	0.823	0.771	0.715	0.588	0.504	0.437	0.382
			0.20	1.000	0.982	0.953	0.919	0.879	0.833	0.782	0.725	0.597	0.511	0.443	0.388
		0.6	0.05	1.000	0.969	0.926	0.881	0.834	0.784	0.732	0.677	0.535	0.459	0.398	0.348
			0.10	1.000	0.974	0.939	0.899	0.856	0.808	0.756	0.701	0.554	0.475	0.412	0.361

(Continued)

TABLE A2.1 (CONTINUED) Stability Factor χ of Axial Compression Member for Circular Concrete-Filled Double Skin Steel Tube

Steel	Concrete	ψ	α_n	λ											
				10	20	30	40	50	60	70	80	90	100	110	120
			0.15	1.000	0.978	0.946	0.910	0.869	0.823	0.771	0.715	0.566	0.485	0.420	0.368
			0.20	1.000	0.981	0.952	0.918	0.878	0.833	0.782	0.725	0.574	0.492	0.426	0.373
		0.75	0.05	1.000	0.965	0.924	0.880	0.833	0.784	0.732	0.677	0.508	0.436	0.378	0.331
			0.10	1.000	0.971	0.937	0.898	0.855	0.808	0.756	0.701	0.554	0.475	0.412	0.361
			0.15	1.000	0.976	0.945	0.909	0.868	0.822	0.771	0.715	0.566	0.485	0.420	0.368
			0.20	1.000	0.979	0.951	0.917	0.878	0.833	0.782	0.725	0.574	0.492	0.426	0.373
S450	C40/50	0	0.05	1.000	0.940	0.878	0.818	0.761	0.705	0.651	0.599	0.517	0.443	0.384	0.336
			0.10	1.000	0.946	0.890	0.835	0.780	0.726	0.673	0.620	0.535	0.459	0.398	0.348
			0.15	1.000	0.950	0.897	0.845	0.792	0.739	0.686	0.633	0.546	0.468	0.406	0.355
			0.20	1.000	0.952	0.902	0.852	0.800	0.748	0.696	0.642	0.554	0.475	0.412	0.360
		0.15	0.05	1.000	0.940	0.878	0.818	0.760	0.705	0.651	0.599	0.514	0.441	0.382	0.334
			0.10	1.000	0.946	0.890	0.835	0.780	0.726	0.673	0.620	0.532	0.456	0.396	0.346
			0.15	1.000	0.950	0.897	0.844	0.792	0.739	0.686	0.633	0.543	0.466	0.404	0.353
			0.20	1.000	0.952	0.902	0.852	0.800	0.748	0.696	0.642	0.551	0.473	0.410	0.358
		0.3	0.05	1.000	0.940	0.878	0.818	0.760	0.705	0.651	0.599	0.506	0.434	0.376	0.329
			0.10	1.000	0.945	0.889	0.834	0.780	0.726	0.673	0.620	0.524	0.449	0.389	0.341
			0.15	1.000	0.949	0.897	0.844	0.792	0.739	0.686	0.633	0.535	0.459	0.397	0.348
			0.20	1.000	0.952	0.902	0.851	0.800	0.748	0.696	0.642	0.543	0.465	0.403	0.353

(Continued)

TABLE A2.1 (CONTINUED) Stability Factor χ of Axial Compression Member for Circular Concrete-Filled Double Skin Steel Tube

Steel	Concrete	ψ	α_n	λ											
				10	20	30	40	50	60	70	80	90	100	110	120
		0.45	0.05	1.000	0.939	0.877	0.818	0.760	0.705	0.651	0.599	0.493	0.423	0.366	0.321
			0.10	1.000	0.944	0.889	0.834	0.780	0.726	0.673	0.620	0.510	0.437	0.379	0.332
			0.15	1.000	0.948	0.896	0.844	0.791	0.739	0.686	0.633	0.521	0.446	0.387	0.339
			0.20	1.000	0.951	0.901	0.851	0.800	0.748	0.696	0.642	0.528	0.453	0.393	0.344
		0.6	0.05	1.000	0.937	0.876	0.817	0.760	0.704	0.651	0.599	0.474	0.406	0.352	0.308
			0.10	0.998	0.942	0.887	0.833	0.779	0.726	0.673	0.620	0.491	0.421	0.365	0.319
			0.15	0.997	0.946	0.895	0.843	0.791	0.739	0.686	0.633	0.501	0.429	0.372	0.326
			0.20	0.998	0.949	0.900	0.850	0.799	0.748	0.695	0.642	0.508	0.436	0.378	0.330
		0.75	0.05	0.995	0.934	0.874	0.815	0.759	0.704	0.651	0.599	0.450	0.386	0.334	0.293
			0.10	0.994	0.939	0.885	0.832	0.778	0.725	0.673	0.620	0.491	0.421	0.365	0.319
			0.15	0.994	0.943	0.893	0.842	0.790	0.738	0.686	0.633	0.501	0.429	0.372	0.326
			0.20	0.994	0.947	0.898	0.849	0.799	0.748	0.695	0.642	0.508	0.436	0.378	0.330
S450	C55/67	0	0.05	0.993	0.916	0.843	0.775	0.712	0.653	0.599	0.550	0.474	0.407	0.352	0.308
			0.10	0.993	0.922	0.855	0.791	0.730	0.673	0.619	0.569	0.491	0.421	0.365	0.319
			0.15	0.993	0.926	0.861	0.800	0.741	0.685	0.631	0.581	0.501	0.430	0.372	0.326
			0.20	0.993	0.928	0.866	0.806	0.749	0.693	0.640	0.589	0.508	0.436	0.378	0.331
		0.15	0.05	0.993	0.916	0.843	0.775	0.712	0.653	0.599	0.550	0.472	0.404	0.351	0.307
			0.10	0.993	0.922	0.854	0.790	0.730	0.673	0.619	0.569	0.488	0.419	0.363	0.318

(Continued)

TABLE A2.1 (CONTINUED) Stability Factor χ of Axial Compression Member for Circular Concrete-Filled Double Skin Steel Tube

Steel	Concrete	ψ	α_n	λ											
				10	20	30	40	50	60	70	80	90	100	110	120
			0.15	0.993	0.926	0.861	0.800	0.741	0.685	0.631	0.581	0.498	0.427	0.370	0.324
			0.20	0.993	0.928	0.866	0.806	0.749	0.693	0.640	0.589	0.506	0.433	0.376	0.329
		0.3	0.05	0.992	0.915	0.843	0.775	0.712	0.653	0.599	0.550	0.464	0.398	0.345	0.302
			0.10	0.992	0.921	0.854	0.790	0.730	0.673	0.619	0.569	0.481	0.412	0.357	0.313
			0.15	0.992	0.925	0.861	0.799	0.741	0.685	0.631	0.581	0.491	0.421	0.365	0.319
			0.20	0.992	0.928	0.866	0.806	0.749	0.693	0.640	0.589	0.498	0.427	0.370	0.324
		0.45	0.05	0.991	0.914	0.842	0.775	0.712	0.653	0.599	0.550	0.452	0.388	0.336	0.294
			0.10	0.991	0.920	0.853	0.790	0.730	0.673	0.619	0.569	0.468	0.401	0.348	0.304
			0.15	0.991	0.924	0.860	0.799	0.740	0.685	0.631	0.581	0.478	0.409	0.355	0.311
			0.20	0.991	0.927	0.865	0.806	0.748	0.693	0.640	0.589	0.485	0.415	0.360	0.315
		0.6	0.05	0.989	0.913	0.841	0.774	0.711	0.653	0.599	0.550	0.435	0.373	0.323	0.283
			0.10	0.988	0.919	0.852	0.789	0.729	0.673	0.619	0.569	0.450	0.386	0.335	0.293
			0.15	0.988	0.922	0.859	0.798	0.740	0.684	0.631	0.581	0.459	0.394	0.341	0.299
			0.20	0.989	0.925	0.864	0.805	0.748	0.693	0.640	0.589	0.466	0.400	0.346	0.303
		0.75	0.05	0.986	0.910	0.839	0.773	0.710	0.653	0.599	0.550	0.413	0.354	0.307	0.269
			0.10	0.985	0.916	0.850	0.788	0.728	0.672	0.619	0.569	0.450	0.386	0.335	0.293
			0.15	0.985	0.920	0.857	0.797	0.739	0.684	0.631	0.581	0.459	0.394	0.341	0.299
			0.20	0.985	0.923	0.862	0.804	0.747	0.693	0.640	0.589	0.466	0.400	0.346	0.303

(Continued)

TABLE A2.1 (CONTINUED) Stability Factor χ of Axial Compression Member for Circular Concrete-Filled Double Skin Steel Tube

Steel	Concrete	ψ	α_n	λ											
				10	20	30	40	50	60	70	80	90	100	110	120
S450	C70/85	0	0.05	0.983	0.897	0.817	0.744	0.677	0.616	0.562	0.514	0.443	0.380	0.330	0.288
			0.10	0.983	0.903	0.828	0.758	0.694	0.635	0.581	0.532	0.459	0.394	0.341	0.299
			0.15	0.984	0.907	0.834	0.767	0.704	0.646	0.592	0.543	0.469	0.402	0.348	0.305
			0.20	0.984	0.909	0.839	0.773	0.711	0.654	0.600	0.551	0.475	0.408	0.353	0.309
		0.15	0.05	0.983	0.897	0.817	0.744	0.677	0.616	0.562	0.514	0.441	0.378	0.328	0.287
			0.10	0.983	0.903	0.828	0.758	0.694	0.635	0.581	0.532	0.457	0.392	0.339	0.297
			0.15	0.984	0.907	0.834	0.767	0.704	0.646	0.592	0.543	0.466	0.400	0.346	0.303
			0.20	0.984	0.909	0.839	0.773	0.711	0.654	0.600	0.551	0.473	0.405	0.351	0.308
		0.3	0.05	0.982	0.896	0.817	0.743	0.676	0.616	0.562	0.514	0.434	0.372	0.323	0.282
			0.10	0.983	0.902	0.828	0.758	0.694	0.634	0.581	0.532	0.450	0.385	0.334	0.292
			0.15	0.983	0.906	0.834	0.767	0.704	0.646	0.592	0.543	0.459	0.393	0.341	0.298
			0.20	0.983	0.909	0.839	0.773	0.711	0.654	0.600	0.551	0.465	0.399	0.346	0.303
		0.45	0.05	0.982	0.896	0.816	0.743	0.676	0.616	0.562	0.514	0.423	0.363	0.314	0.275
			0.10	0.982	0.902	0.827	0.758	0.693	0.634	0.581	0.532	0.438	0.375	0.325	0.285
			0.15	0.982	0.905	0.833	0.766	0.704	0.646	0.592	0.543	0.447	0.383	0.332	0.291
			0.20	0.982	0.908	0.838	0.773	0.711	0.654	0.600	0.551	0.453	0.389	0.337	0.295
		0.6	0.05	0.980	0.895	0.815	0.743	0.676	0.616	0.562	0.514	0.407	0.349	0.302	0.265
			0.10	0.980	0.900	0.826	0.757	0.693	0.634	0.581	0.532	0.421	0.361	0.313	0.274

(Continued)

TABLE A2.1 (CONTINUED)   Stability Factor χ of Axial Compression Member for Circular Concrete-Filled Double Skin Steel Tube

Steel	Concrete	ψ	α_n	λ											
				10	20	30	40	50	60	70	80	90	100	110	120
			0.15	0.980	0.904	0.833	0.766	0.703	0.645	0.592	0.543	0.430	0.368	0.319	0.279
			0.20	0.980	0.907	0.837	0.772	0.711	0.653	0.600	0.551	0.436	0.374	0.324	0.284
		0.75	0.05	0.977	0.892	0.814	0.742	0.675	0.616	0.562	0.514	0.386	0.331	0.287	0.251
			0.10	0.976	0.898	0.824	0.756	0.692	0.634	0.581	0.532	0.421	0.361	0.313	0.274
			0.15	0.977	0.902	0.831	0.764	0.703	0.645	0.592	0.543	0.430	0.368	0.319	0.279
			0.20	0.977	0.905	0.836	0.771	0.710	0.653	0.600	0.551	0.436	0.374	0.324	0.284

PS: The intermediate values can be calculated by linear interpolation.

where

χ is the stability factor of axial compression member

ψ is the hollow section ratio

α_n is the nominal steel ratio

λ is the slenderness ratio

Appendix 3: Examples for the Bearing Capacity Calculation of CFDST Members

Basic conditions:

Cross-Sectional Size and Load Information

Outer Tube[a] $D_o \times t_o$ (mm)	Inner Tube[a] $D_i \times t_i$ (mm)	Effective Length (m)	Steel Grade	Concrete Grade	Design Value of the Internal Force	
					Axial Force N (kN)	Moment M (kN·m)
1000×22	500×18	4.8	S355	C40/50	12000	8000

[a] Steel tubes are hot rolled structural steel.

Geometrical information:

$$D_o = 1000\,\text{mm},\ t_o = 22\,\text{mm},\ D_i = 500\,\text{mm},\ t_i = 18\,\text{mm},\ L_e = 4800\,\text{mm}$$

$$A_{si} = \pi \times [500^2 - (500 - 36)^2]\,/\,4 = 2.73 \times 10^4\,\text{mm}^2$$

$$A_{so} = \pi \times [1000^2 - (1000-44)^2]/4 = 6.76 \times 10^4 \, mm^2$$

$$A_c = \pi \times ((1000-44)^2 - 500)^2 /4 = 5.21 \times 10^5 \, mm^2$$

$$A_{ce} = \pi \times (1000 - 2 \times 22)^2 /4 = 7.18 \times 10^5 \, mm^2$$

$$A_{sc} = A_{so} + A_c + A_{si} = 6.16 \times 10^5 \, mm^2$$

Physical information:

According to EN 1992-1-1 (Section 3.1) and EN 1993-1-1 (Section 3.2)

$$E_s = 2.1 \times 10^5 \, N/mm^2$$

$$E_c = 3.5 \times 10^4 \, N/mm^2$$

$$f = f_y = 355 \, N/mm^2$$

$$f_u = 510 \, N/mm^2$$

$$f_{ck} = 40 \, N/mm^2$$

$$f_c = f_{ck}/\gamma_0 = 40/1.5 = 26.7 \, N/mm^2$$

a. Axial compressive capacity

$$\alpha = A_{so}/A_c = 0.13, \quad \alpha_n = A_{so}/A_{ce} = 0.094$$

$$\xi = \alpha_n \cdot f_{y,outer}/f_{ck} = 0.094 \cdot 355/40 = 0.8343$$

$$\xi_o = \alpha_n \cdot f_{outer}/f_c = 0.094 \times 355/26.7 = 1.2498$$

$$\psi = \frac{D_i}{D_o - 2t_o} = 0.523$$

$$C_1 = \alpha/(1+\alpha) = 0.115, \quad C_2 = (1+\alpha_n)/(1+\alpha) = 0.968$$

$$f_{osc} = C_1 \cdot \psi^2 \cdot f_{yo} + C_2 \cdot (1.14 + 1.02\xi_o) \cdot f_c$$

$$= 0.115 \times 0.523^2 \times 355 + 0.968 \times (1.14 + 1.02 \times 1.2498) \times 26.7 = 73.6 \,\text{N/mm}^2$$

$$N_{osc,u} = f_{osc} \cdot (A_{so} + A_c) = 73.6 \times 5.89 \times 10^5 \times 10^{-3} = 4.33 \times 10^4 \,\text{kN}$$

$$N_{i,u} = f_{inner} \cdot A_{si} = 355 \times 2.73 \times 10^4 \times 10^{-3} = 9.69 \times 10^3 \,\text{kN}$$

$$N_u = N_{osc,u} + N_{i,u} = 5.299 \times 10^4 \,\text{kN}$$

$$I = \pi \cdot [D_o^4 - (D_i - 2t_i)^4]/64 = 4.602 \times 10^{10} \,\text{mm}^4$$

$$r = \sqrt{I/A_{sc}} = \sqrt{4.602 \times 10^{10}/616000} = 273.3 \,\text{mm}$$

$$\lambda = L_e/r = 4800/273.3 = 17.56$$

The corresponding stability factor can be obtained according to Appendix 2

$$\chi = 0.956$$

b. Flexural capacity

$$\gamma_{m1} = 0.48\ln(\xi + 0.1)(1 + 0.06\psi - 0.85\psi^2) + 1.1 = 1.074$$

$$\gamma_{m2} = -0.02\psi^{-2.76}\ln(\xi) + 1.04\psi^{-0.67} = 1.627$$

$$W_{scm} = \frac{\pi \cdot (D_o^4 - D_i^4)}{32D_o} = 9.20 \times 10^7 \,\text{mm}^3$$

$$W_{si} = \frac{\pi[D_i^4 - (D_i - 2t_i)^4]}{32D_i} = 3.17 \times 10^6 \,\text{mm}^3$$

$$M_{osc,u} = \gamma_{m1} \cdot W_{scm} \cdot f_{osc} = 1.074 \times 9.20 \times 10^7 \times 72.9 \times 10^{-6} = 7.2 \times 10^3 \,\text{kN} \cdot \text{m}$$

$$M_{i,u} = \gamma_{m2} \cdot W_{si} \cdot f_{y,inner} = 1.627 \times 3.17 \times 10^6 \times 355 \times 10^{-6} = 1.83 \times 10^3 \text{ kN} \cdot \text{m}$$

$$M_u = M_{osc,u} + M_{i,u} = 9.03 \times 10^3 \text{ kN} \cdot \text{m}$$

c. Check for requirements

$$N_E = \frac{\pi^2 \cdot (EI)_{sc}}{L_e^2} = 1.367 \times 10^6 \text{ kN}$$

$$\zeta_o = 1 + (0.18 - 0.2\psi^2)\xi^{-1.15} = 1.154$$

$$\xi = 0.8343 > 0.4$$

$$\eta_o = (0.1 + 0.14 \cdot \xi^{-0.84}) \cdot (1 + 0.7\psi - 1.8\psi^2) = 0.2424$$

$$b = \frac{1 - \zeta_o}{\chi^3 \cdot \eta_o^2} = -3.135; \quad c = \frac{2 \cdot (\zeta_o - 1)}{\eta_o} = 1.27; \quad d = 1 - 0.4 \cdot (\frac{N}{N_E}) = 0.9965$$

$$\frac{N}{N_u} = \frac{1.2 \times 10^4}{5.299 \times 10^4} = 0.226 < 2\chi^3\eta_o = 2 \times 0.956^3 \times 0.2424 = 0.424$$

According to 1994-1-1 (Table 6.4), in this case $\beta_m = 1.0$

$$-b \cdot \left(\frac{N}{N_u}\right)^2 - c \cdot \left(\frac{N}{N_u}\right) + \frac{1}{d} \cdot \left(\frac{M}{M_u}\right) = 0.791 \leq 1$$

Therefore, this column is safe to be used in practice.

Index

Printed in the United States
by Baker & Taylor Publisher Services